Beyond the Large Farm

Beyond the Large Farm

Ethics and Research Goals for Agriculture

EDITED BY

Paul B. Thompson
and Bill A. Stout

Routledge
Taylor & Francis Group

LONDON AND NEW YORK

First published 1991 by Westview Press, Inc.

Published 2018 by Routledge
52 Vanderbilt Avenue, New York, NY 10017
2 Park Square, Milton Park, Abingdon, Oxon OX14 4RN

Routledge is an imprint of the Taylor & Francis Group, an informa business

Library of Congress Cataloging-in-Publication Data
Beyond the large farm : ethics and research goals for agriculture /
Paul B. Thompson and Bill A. Stout, editors.
   p. cm.
Includes bibliographical references and index.
ISBN 0-8133-8192-4
   1. Agriculture—Research—United States—Moral and ethical
aspects. 2. Agriculture—Government policy—United States.
3. Agriculture—Research—Moral and ethical aspects. I. Thompson,
Paul B., 1951-   . II. Stout, B. A.
S541.B48 1991
174'.936—dc20   90-20977
                                                                  CIP

ISBN 13: 978-0-367-01624-1 (hbk)
ISBN 13: 978-0-367-16611-3 (pbk)

Dedicated to
Dr. H.O. Kunkel

Contents

PART TWO
Bibliographic Essays on the Critics of Agriculture

Preface

In editing this book, we have thought of our contributed chapters as developing an argument which begins by examining the rationale for emphasizing productivity as the dominant goal of agricultural research, and which then examines challenges in the form of alternative goals that scientists might seek in performing agricultural research. These chapters are followed by bibliographic essays that review some of the criticisms of agricultural research during the last 2 decades. The point of these chapters is to place the main argument of the book—a review of how scientists think of research goals—into a broader science policy context, in which the voices of non-scientist critics are important.

This general plan required a series of editorial decisions that were somewhat arbitrary. First was the question of how to arrange the chapters themselves. We wanted the book to read with an argumentative structure, rather than like a typical anthology's collection of papers. We finally chose an order that seemed to make the most sense from this perspective. That choice meant separating bibliographic essays into a second part, in which their supporting role is more apparent; but it would be just as logical to read them before launching into the main argument of the book.

Second, we finally decided to organize the book with a master bibliography, rather than with separate bibliographies following each chapter. The master bibliography promised to be of the most use to our readers, and it eliminated the need to reprint the same bibliographic item two, three, and even four times. At the same time, creation of a bibliography is itself a part of the scholarly research process, and our choice makes it much more difficult to see which segment of the literature has influenced each author. This is particularly problematic for the bibliographic essays, for which a main part of the work was in collecting the citations themselves. Our choice makes the work of our bibliographers less apparent to the reader, and for that, we apologize.

Paul B. Thompson
Bill A. Stout

Acknowledgments

Books such as this one require a great deal of dedication and work behind the scenes. The *Beyond the Large Farm* project at Texas A&M University was funded by the Ethics and Values Studies Program at the National Science Foundation. We wish to thank those at NSF and elsewhere in government who have seen fit to create such programs, and especially Rachelle Hollander, whose advice on the project was invaluable.

The project involved monthly seminars attended faithfully by a "core group" of scientists from Texas A&M University. The group contributed to the structure and choice of authors for this book, and offered detailed criticisms and suggestions on drafts of the respective chapters. They were: Katherine Dettwyler, Richard Frederickson, Marvin Harris, H. O. Kunkel, E. Wesley F. Peterson, and Don Vietor. Two authors for the volume, G. Lain Ellis and Deborah Tolman, also participated in the core group and helped with the organization and coordination of its activities. Others who made special contributions to the success of the project were Norman Borlaug, Neville Clarke, and James Stack.

All the chapters for this book are appearing here for the first time with the exception of the one by Vernon Ruttan, which is a significantly (if subtly) revised version of material published in Ruttan (1982a) and as "Moral Responsibility in Agricultural Research," *Southern Journal of Agricultural Economics* 15 (1983) 73-80.

Terri Huff was the technical editor for the volume, and her skills and expertise are largely responsible for the continuity and high quality of the text. Jerry Arnett was the key person responsible for keeping track of each author's manuscript, and for ensuring that the text was accurately transcribed. Rosalyn Chatham and Bradley Rickleman helped with the bibliography during the early stages of manuscript preparation. In the final stages, Doug Kutach and Kelly Hancock did yeoman's service in bringing the manuscript to its final form. Responsibility for remaining errors and flaws is, of course, entirely our own.

P.B.T.
B.A.S.

PART ONE

Ethics and Research Goals
for Agriculture

Introduction

Values in the Agricultural Laboratory

Paul B. Thompson, G. Lain Ellis, and Bill A. Stout

Overview

In 1987, U.S. agricultural research concluded 100 years of administration and funding under the Hatch Act. This law established the USDA/land-grant triad of government/experiment station/university, and supported science and technology for agriculture through non-competitive, centrally administered public funds. American agriculture has enjoyed a reputation for success during much of its tenure, but in the last 20 years, agricultural research priorities have been subjected to criticism and reevaluation (Johnson, 1984; Danbom, 1986; Kirkendall, 1987). One important group of critics, associated with the 1972 "Pound" Committee of the National Research Council (NRC) (NRC, 1972; 1975), stressed the scientific quality and efficiency of agricultural research. However, the more celebrated critics have focused upon the social goals that contemporary agricultural research have tended to serve (whether implicitly or intentionally) (Hightower, 1975; Berry, 1977; Jackson, 1980; Schell, 1984; Doyle, 1985; Fox, 1986).

This latter group of critics has expressed an extraordinary range of concerns and complaints, including the allegation that many different client groups have been ill-served. An element in most criticisms, however, is that agricultural research leaders have, *de facto* or by design, pursued the goal of maximizing the productive efficiency of large farms, while ignoring other research goals. Critics allege that the USDA/land-grant system's persistent search for greater yields is the wellspring of problems for American agriculture.

The views of these critics were reinforced by a recent legal finding.

In November 1987, California Rural Legal Assistance (CRLA) won a judgment against the University of California (UC). CRLA claimed that producers who aggressively sought a competitive edge benefited disproportionately from publicly funded agricultural science at the expense of small farms and farm labor that were displaced by resulting technological changes. The court found that UC negligently failed to assess whether research to develop a mechanical tomato harvester (among other biological and mechanical technologies) would have an adverse impact upon UC's legislatively mandated small farm clients (Bishop, 1987; Hager, 1987; Sinclair, 1987). The judge traced the university's legal commitment to serve small farms back through a series of appropriations bills, all the way to the original Hatch Act of 1887. Other critics cite negative impacts on environmental quality, the poor or oppressed citizens of developing nations, or consumer health. But the common theme is that the USDA/land-grant's service to increasingly large farms has lowered the price and improved the availability of food and fiber at the expense of other social goals.

The technical accuracy of this charge has been challenged on several fronts[1], and examining the legitimacy of these critical opinions is a continuing need. Indeed, the decision in the CRLA/UC case was overturned by a California appellate court's ruling that UC did not violate its mandate because service to small farm clients is not a specific mission of the experiment station. The reversal of this decision illustrates that the mission of the research system is open to widely divergent interpretation. Moreover, many scientists and research administrators within the USDA/land-grant system have accepted the critics' suggestion that the research agenda for agriculture must be broadened to reflect a wider range of social goals, (Ruttan, 1983; Jordan et al., 1986). Therefore, we must move beyond the negative critique of past agricultural research and practice, and begin establishing a vision of the goals that might guide agriculture in the future.

In this chapter, we will suggest that agricultural research policy can be characterized aptly in terms of establishing and developing the components of the research laboratory. We will then present arguments supporting three points.

1. Implicit ethical goals have always existed within American agricultural research in the USDA/land-grant system. Although these goals are rarely articulated or defended, they nonetheless exist, and, in fact, have given agricultural research policy a firm foundation in one of the enduring traditions of ethics for the modern age.
2. American agricultural research institutions have consistently relied

on a specific pattern of norms and procedures to achieve the implicit goal. Although these norms and procedures are also implicit, they have, from time to time, been an object of discussion relative to agricultural research policy. They can be analyzed as a strategy for achieving increased farm productivity. The strategy is consistent with (and related historically to) the ethical tradition that shapes the implicit goal of agricultural research.

3. When we analyze the ethical foundation of the goal and the strategy for achieving it, we discover that many of the criticisms raised against the agricultural research system are consistent with the traditional ethical goals of American agricultural research, at least at the foundational level. To the extent that this is true, a mere modification of traditional goals is needed, rather than a full-scale attack upon them.

Finally, we will introduce the contributed chapters in this volume. The book itself is the end product of a project initiated by the editors, Paul Thompson and Bill Stout. The purpose of the project, which was supported by grants from the National Science Foundation (NSF) Ethics and Values Studies Program, was to establish dialogue between agricultural scientists, humanists, and social scientists who share an interest in the ethical and social aspects of agricultural research policy. The project involved a group of Texas A&M University faculty who met monthly to discuss issues in agricultural research policy. This "core group" included representatives from the departments of agricultural economics, agricultural engineering, animal science, anthropology, entomology, philosophy, plant pathology and microbiology, and soil and crop sciences.

The group conducted a series of four workshops in which the project members and guest speakers, together with other agricultural and social scientists, discussed some of the central issues in current agricultural research policy. The contributed chapters are the result of those workshops, and represent the guest speakers' attempts to resolve some of the conflict that has plagued agriculture during the last 2 decades. Even if the contributing authors have made some progress toward making our agriculture more responsive to a broader set of ethical goals, it is presumptuous to suggest that fundamental conflicts in agricultural research have been resolved here in any final sense. The suggestions offered by the authors tend to accept the traditional foundation of agricultural research and offer strategies for adaptation and reform. Therefore, the background of this book assumes that the ethical foundations for agricultural research are still valid, but the application of these traditional goals to specific research choices can be modified to accommodate valid criticisms. By contrast, some critics of agricultural research pose more radical and more fundamental critiques. The three bibliographic essays in this book survey some of this critical literature, but these critics probably would not be satisfied by the reforms proposed herein. The authors maintain that the current system can be significantly improved, an assumption we will examine in this chapter.

The Laboratory Concept of Agricultural Research Policy

It is useful to think about agricultural research policy in terms of the establishment or modification of a laboratory that performs certain kinds of activities. In its narrowest sense, the laboratory is a collection of equipment, materials, and physical amenities with which to perform research. More broadly, the laboratory is the "equipped space" in which a scientist produces or tests theories, facts, and technologies. The "equipped space" can be an agronomist's field plots as easily as it can be a room filled with scientific machinery and supplies, our usual idea of a laboratory. In an even broader sense, however, the laboratory includes people. The laboratory concept is helpful in this context because it suggests a natural linkage between research capacity, research goals, and the products of scientific research. In the broadest sense, the laboratory includes its operators' goals, values, and theoretical approaches as part of its equipment because these abstract components structure the ways in which people use the physical laboratory (Latour, 1987). Goals and theories are important elements of the laboratory for distinctive, but related reasons.

Theories are a product of scientific research, but theoretical approaches also define the parameters within which problems are framed and solutions proposed. This, in turn, affects how the laboratory is equipped and reequipped during the course of research. Setting up the physical equipment in a laboratory enables the researcher to perform certain kinds of activity, but it also precludes pursuing other kinds of activities: A laboratory designed to observe subcellular or molecular phenomena may have little capacity for doing research at the organismic or ecosystem level. In a similar way, equipping a laboratory with scientists who are committed to one disciplinary approach makes it highly unlikely that the laboratory will produce an advance in some other area. There is, thus, a double sense in which theory structures human activity within the laboratory. First, theory defines the program of scientific objectives (or goals) that can be expected to advance the discipline. Theories are often being tested and produced in the laboratory setting. Moreover, a given theoretical perspective defines what researchers will recognize as good science. Second, theory dictates the purchase of equipment, which, in turn, dictates limits upon what can be done in the laboratory even when theories undergo radical changes.

Goals have a profound (if subtle) impact on research precisely because research is a goal-oriented activity. (It takes little imagination to predict the outcome of a funding proposal that has no goals.) However, the manner in which specific research goals are defined is affected by other, more general goals. For example, the goal of perfecting a research technique for monitoring a plant's intake of a given nutrient may follow from the more general goal of developing cost-effective fertilizer practices, which may follow from the still more general goal of increasing the profitability of the plant under commercial production conditions. "Monitoring nutrient intake," and "developing cost-effective fertilizer," are importantly different types of goals for scientific research. Conventionally, the difference might be described as follows: Monitoring nutrients is a scientific goal, while developing fertilizer is a goal for application of the science. While this conventional way of describing goals is unexceptional in most instances, it may imply that the goals of application are entirely separate from the goals of science proper. In point of fact, the ability to separate research and application varies considerably from one type of scientific activity to another. Scientists in the Texas A&M University core group expressed the opinion that research involving the transfer of genetic material from one organism to another obliterates the distinction between basic and applied research, because researchers will be attempting to initiate the transfer of traits that are known to be valuable for specific applications.

The laboratory metaphor is a useful way to reconceptualize the basic/applied distinction as it relates to the goals of scientific activity. The laboratory is a place in which research occurs. Some goals are incorporated into the laboratory itself, either because they are specific research goals sought by laboratory personnel, or because the equipment of the lab structures human activity toward one set of results, rather than another. Other goals are outside the laboratory, including the institution's reason for building the laboratory, and most especially society's broad expectations for using research products in a particular way. The inside/outside method of characterizing how goals affect scientific activity is not fully precise, but it does distinguish between goals that are sought as part of the laboratory's research activity, and the myriad of other goals and interests that affect (and are affected by) science. Funding and administrative forces influence the laboratory from the outside, and they can significantly affect what kinds of laboratories are constructed. Once constructed and staffed, however, the laboratory must set internal goals that are usually described in terms of verification, measurement, theoretical development, and methodological advancement.

The problem facing the USDA/land-grant system is one of tension between inside and outside goals. Agricultural research laboratories have invested billions of dollars in equipment and buildings. More importantly, they have invested a century of training and human activity into the development of theories, the collection of data, and the creation of research techniques, all organized around the general goal of increasing the biological productivity of domesticated plants and animals. The history and philosophy of agricultural disciplines, not to mention the specific individuals who staff USDA/land-grant research facilities, set goals within the laboratories of agricultural science. Society as a whole has settled into social patterns, whereby the USDA/land-grant system is expected to further the social goals of agriculture. While these goals have changed throughout time (Thompson, 1986), there has been a serviceable fit between the internal goals of agricultural laboratories and the external goals and expectations of society at large. A new set of general, practical goals may require a very different research program. There is no guarantee, however, that the capacity of a given laboratory, determined by its equipment, personnel, and the theoretical vision within it, will be able to modify its activity in response to new, alternative goals. The potential for tension between laboratory capacity and the more extensive goals that research is assumed to serve sets the stage for a philosophical debate that is addressed more completely by the contributing authors of this book.

However, in adopting (or complying with) a set of goals, a researcher is also adopting (or complying with) a set of values that establish the priority of the chosen goals over other goals that were not chosen. Thus, the laboratory concept of research policy involves the notion that establishing and developing a research stream requires procedures that connect the scientific capability of the laboratory and its internal goals to a set of values. Before proceeding to the chapters on the ethical justification of research goals, we need to obtain some perspective on how disciplinary alliances, theoretical commitments, and beliefs about the links between research and social goals influence the research choices of scientists. The seminal work of Lawrence Busch and William Lacy forms a basis for analyzing problem selection by agricultural scientists.

The Busch and Lacy Study

In *Science, Agriculture, and the Politics of Research*, Busch and Lacy (1983) document how agricultural research is influenced by disciplinary values and the goal of productivity. Busch and Lacy claim that the evolution of agricultural science has been marked by several trends. In the early stages of USDA/land-grant development, science was considered widely as the key to solving social problems. Within the context of this shared belief, agricultural researchers strove for status and prestige equal to other scientists. In the drive toward professionalization, agricultural research became increasingly oriented around disciplinary themes and organized in disciplinary departments. Simultaneously, an increasingly wide rift emerged between basic and applied science, and between research and extension (Schweikhardt and Bonnen, 1986). Busch and Lacy note that the original emphasis upon social engineering as a goal of research has remained important, but agricultural research has been increasingly influenced by its corporate structure. At first, agricultural scientists were encouraged to pursue goals that were validated by research colleagues through the mechanisms of tenure, promotion, and prestige. More recently, peer review has come to play a more direct role in problem choice. As formula funding levels have stagnated relative to the growth in research needs and costs, scientists have been forced to seek competitive grants for research support.

Busch and Lacy's study is based on an analysis of questionnaires returned by 1,431 randomly selected agricultural scientists. Much of the study is based on the respondents' rankings of 21 criteria that influence their choice of research projects. Overall, the scientist's personal interests, availability of facilities, scientific curiosity, and assessment of client needs are highly ranked criteria for problem choice. Scientists report that other researchers in the field, institutional priorities, clientele, colleagues, and extension personnel are relatively uninfluential in problem selection. These findings reinforce the laboratory analysis of research policy. Of the four highly ranked criteria, three (personal interests, availability of facilities, and scientific curiosity) figure in the equipment of the laboratory as characterized previously. The fourth criterion, assessment of client needs, refers to the way specific research goals relate to broader and more general social goals, justifying the existence of the laboratory in a broader social context.

If Busch and Lacy's research has been interpreted correctly, scientists perceive themselves to be more influenced by abstract values than by specific individuals when choosing research projects. For example, client needs, as assessed by the scientist, ranks relatively high (seventh of 21, respectively) as a perceived influence on problem choice, while demands raised by clientele and feedback from extension rank relatively low (thirteenth and twentieth of 21). While scientists resist the influence of external pressure, they equip their laboratories with values that determine which topics are of interest, which topics stimulate their curiosity, and which research results serve the needs of others. These abstract components of the laboratory impose a structure that affects what happens there, just as do the machinery and supplies with which it is equipped.

Busch and Lacy's study shows that, in general, disciplinary structure appears to have a significant impact on research. This structure is most evident in the scientist's publication activities and perceptions of criteria for promotion. Professional journals serve as the primary outlet for research results. Most scientists publish in journals within their own disciplines, with 60 percent of them publishing in three or fewer journals, but only 8 percent publishing in more than six journals. Second, journals constitute the most important published research resource. Most scientists subscribe to three or fewer journals in their own or closely related fields. Only nutritionists and geneticists subscribe to journals in an average of more than three fields. Third, the journals themselves are often highly specialized and may not be accessible to researchers in other fields. Moreover, they are usually published by disciplinary societies with editing and peer review preferred by members of the discipline.

Busch and Lacy believe that the pattern of journal publication and use reflects a high degree of isolation between researchers in different disciplines as well as between researchers and their clientele. Publication in journals has become increasingly important, while other media have stagnated or declined in importance. The lack of importance placed on publication in bulletins, reports, and farm/industry journals for problem selection and promotion criteria, together with the perceived minor influence of clients and extension personnel, suggest that the gap between researchers and their constituencies is wide indeed. The diversity in publication and subscription suggests that scientists in different disciplines have relatively little input into each other's research.

Influences Within and Upon the Laboratory

Indeed, scientists perceive the publication of many journal articles to be the most important criterion for tenure and promotion. This perception apparently contradicts the perception of scientists that evaluation by other scientists in one's field (tenth of 21) and the credibility of others conducting similar research (fourteenth of 21) are relatively unimportant problem selection criteria. In the peer review process, other scientists are the people most likely to determine a researcher's publication rate, and hence, to affect promotion and tenure. Consequently, say Busch and Lacy, (1983:79), the growing role of journal publication "reflects a shift toward the disciplinary association in the power to (1) validate and certify research findings, and (2) less directly, to reward scientists." This result is quite consistent with the laboratory analysis of agricultural research policy. The values that determine the boundaries of interest and scientific curiosity are inculcated in the disciplinary life of scientists—in their training, and participation in a department's or professional organization's activities. But they also are carried *into* the laboratory and form a component of its capacity. Other scientists, even those in the researcher's own discipline, are outside the laboratory. They may influence research problem choice by offering suggestions, promising favors, or, in some cases, exerting power, but such forms of influence might be construed as compromising the laboratory's self-direction and autonomy in setting goals. The scientists' reports reflect the way that disciplinary values structure a laboratory's activities from within, rather than as an external force. As internal values, disciplinary standards are accepted as legitimate principles for structuring research.

More obvious external influences are not absent from the scientists' reports, however. Funding is a fairly highly ranked criterion for problem selection (ninth of 21), as it is for promotion and tenure (third of 8). Many scientists believe they spend too much time and effort trying to acquire funding, which also appears to involve some compromises indicated by the comments of some scientists, quoted anonymously (Busch & Lacy, 1983:162-163):

> Funding, the availability of it, and who is the granting agency determines the kind of research that will be done.

> The general decline in hard funds forces too many to seek...grants (NSF, NIH, commodity, etc.). This in many cases places the need for funding in a higher priority than where research needs are greatest—in short, it controls research directions too much.

Busch and Lacy's data suggest that funding agencies exert influence *upon* the research laboratory, rather than within it. Funds from the USDA, foundations, and state governments are distributed relatively evenly across disciplines, although foundation grants are concentrated in prestigious institutions. Grants from the NSF and National Institutes of Health (NIH) tend to be concentrated in the basic sciences and nutrition and in more prestigious universities, however. Projects funded by these latter agencies often have little or no agricultural importance, and the research agendas at prestigious universities tend to downplay applied research. Thus, the increasing importance of competitive grants represents a new form of external influence upon problem choice in the laboratory. Moreover, the source of funding has several relationships to problem choice. NSF and NIH grant recipients tend to emphasize criteria related to the advancement of disciplinary knowledge, and to deemphasize feedback from clients and extension. Recipients of corporation and commodity association grants tend to emphasize utility-related criteria and feedback from clients and extension in problem choice. Furthermore, high levels of consultation with private corporations can introduce potential conflicts of interest.

Busch and Lacy claim (1983:163) that "it is likely that in almost all cases: (1) scientists choose potential granting organizations according to their interests and the probability of receipt of funds, and (2) those granting organizations have an influence on the scientists." The competition for funds inhibits long-term research because obtaining long-term funding is difficult, and it uses time and effort that could be devoted to research. Moreover, the availability of funding for a given topic "is, in part, the result of overtly political actions [by] both elected officials and various interest groups. These actions, as well as the actions of scientists, determine who benefits from agricultural research" (Busch & Lacy, 1983:165).

As criteria for problem choice, the high rankings of importance to society (second of 21) and client needs (albeit, assessed by the scientist) suggest that agricultural scientists take seriously the need to justify their activities with respect to broader goals. Busch and Lacy's data indicate that scientists perceive the actual and preferred beneficiaries of research to be (in descending order) the general public, large farmers, other disciplines, small farmers, agribusiness, federal agencies, rural residents, state and local agencies, and foreign groups. Their data also show that scientists from non-farm backgrounds see small and large farmers, agribusiness, and rural residents (the traditional target groups) as less likely to benefit from research. Of interest, scientists from non-farm backgrounds rank scientific criteria for problem choice higher than utility-related criteria, while the reverse is true for scientists from farm backgrounds, who as a group tend to be older.

Busch and Lacy suggest that as the farm population shrinks and future scientists are increasingly recruited from urban areas, scientists likely will become increasingly detached from farmers' problems (MacKenzie, this volume), and will increasingly consider the advancement of disciplinary science as an important goal of research. This trend is already apparent: Out of 11 goals for research, the most highly ranked is development of new knowledge and improved methodology, followed closely by increased productivity. However, the relatively low rankings of the goals improving rural level of living (sixth), assisting developing nations (eighth), and promoting community improvement (ninth) suggest that importance to society as a criterion for problem choice, "glosses over a rather heterogeneous group of meanings" (Busch & Lacy, 1983:44).

Furthermore, Busch and Lacy note (1983:233) that there is "a surprising lack of concern for contemporary agricultural issues among agricultural scientists, particularly among basic scientists and those with a basic research orientation." Scientists rated their perception of the degree to which they perceive their research to reflect five emerging issues (on a scale of 1="rarely" to 7="frequently"). As a group, scientists perceive only environmental issues (mean = 4.4) to be reflected above the midpoint of the scale in their research, followed by the world food crisis (mean = 3.6), energy issues (3.5), and consumer issues (3.2). Scientists believe that their research reflects alternative approaches to agriculture only to a minimal extent (mean = 2.6).

It is tempting to speculate that this ranking, which Busch and Lacy relate to the agricultural scientists' "concern for contemporary agricultural issues," really reflects the level of research capacity (including technology, personnel, and values) that existed within agricultural research laboratories at the time Busch and Lacy published their work. That is, scientists perceived themselves to have little ability to align their research activities in concordance with the broad goals implied by each of these issue areas. Different reasons may be cited for such a judgment, ranging from a lack of imagination about how research activities might serve these goals, to a philosophical opposition to the goals themselves.

Thus, based on Busch and Lacy's account, the research agenda setting process is highly complex, and reflects a number of constraints. Many scientists select problems that they perceive to be important for society, consistent with the goal of serving the public interest. On the other hand, disciplinary influences and funding pressures appear to narrow the range and scope of problems that can be addressed, especially in relation to alternative agricultural strategies. Quoted in Busch and Lacy (1983:213), an anonymous university scientist suggests, "I don't believe that anyone interested in small or organic farming could get tenure here as he or she would be considered off the beaten track." Busch and Lacy believe that the reward and funding systems must be altered to promote high-risk, long-term research, and that efforts must be made to increase interdisciplinary contact and research to promote creativity and innovation. Although Busch and Lacy do not use the laboratory concept to analyze agricultural research, their findings provide a basis for understanding how values and loyalties function within the confines of the laboratory, as well as how other forces form an external influence on it.

Productivity and Utilitarianism

What Busch and Lacy do not provide is any account of why the values of agricultural scientists would lead them to configure research laboratories to increase agricultural productivity, as opposed to an alternative goal. A philosophical study of agriculture should provide some basis for understanding the traditional view of how production decisions might be justified in light of social and ethical goals. Although articulating and defending any full vision of the appropriate philosophy for American agriculture itself is far beyond the scope of this book, we can describe the philosophical framework that appears to shape the reasoning of many agricultural scientists and administrators. Because agricultural research is intended to improve the practice of agriculture, this framework is most relevant to research choices. If we take our cue from the common theme of agriculture's critics, increasing farm productivity is the goal that has guided agriculture during the past century. Although the defense of this goal may be obvious to many, it is worth documenting not only its logic, but also its pedigree.

The philosophical rationale for increasing agricultural productivity forms an explicit part of John Locke's 1690 argument on private property:

> ...he that encloses land, and has a greater plenty of the conveniences of life from ten acres, than he could have from an hundred left to nature, may truly be said to give ninety acres to mankind: for his labour now supplies him with provisions out of ten acres, which were but the product of an hundred lying in common (Locke, 1690, pp 23-24).

Locke's idea is that the multiplication of benefits (the conveniences of life) that follow an increase in the productivity of agricultural land serves as a moral justification of social policies (here, the establishment of private property) that achieve this end. The ethical justification for this viewpoint is neither complicated nor even controversial. If more food and fiber is produced, there will be more to go around; hence, a clear benefit to the consumers of agricultural products. If food and fiber can be produced more efficiently, the producers will have more time and money for enjoying the other "conveniences" of life. If other things are equal, any possibility of improving the productive capacity and efficiency of agriculture is a good thing.

This form of philosophical reasoning stresses the comparison of costs and benefits arising from taking a given course of action. It was offered as an explicit moral framework for the evaluation of public policy in Jeremy Bentham's 1789 *Principles of Morals and Legislation*. Ethical theories that evaluate actions by comparing the net benefits of all options have been dubbed "utilitarian" ever since, and utilitarian moral standards have been explicitly applied to the food and farm system in recent discussions (Tweeten, 1983; Thompson, 1988). To some people, Bentham's moral calculus seems to be little more than common sense. To determine what one ought to do, one simply compares the benefits and harms of each option. The choice that leads to the greatest net benefit (or the least net harm) is the choice that is morally right. The emphasis upon maximization or optimization of benefits is what we mean by the *utilitarian maxim.*

Although Locke is not typically regarded as a utilitarian philosopher, the utilitarian maxim is consistent with the argument he uses to justify enclosing land. He might be described as utilitarian in the sense that the change in agricultural land holdings that he advocates would be justified if it optimized (or at least improved) the benefits traditionally derived from agriculture. The argument could be applied just as easily to a new feeding regimen for livestock, a new chemical, or even an investment in public education. Three features of this argument should be noted for future reference. First, the utilitarian maxim applies equally to all people affected by an action. The justification of an agricultural technology may depend far more on the benefits to consumers in the form of reduced food costs than on benefits for producers themselves. Second, the utilitarian maxim recognizes that some may not benefit and, indeed, may be harmed by the action. It was certainly Locke's understanding that enclosure would threaten the farming practices of his day because it would eliminate the commons around which agriculture was structured. The argument is that although some are harmed, others benefit to a greater degree, so net benefits exceed net harms. This means that the optimizing strategy of a utilitarian argument is one that views ethics as the theory of making trade-offs— balancing benefits for some against harm for others.

Third, the range of utilities (benefits) includes not only utilities that are desirable as direct benefits, but also utilities that are desirable because of their capacity to favorably affect future outcomes. In John Stuart Mill's *Utilitarianism* (1861), this latter notion of utility plays two roles. Mill held intellectual development in very high regard. It is desirable (i.e., it has utility) because it fosters a sense of enlightened self-interest. Well-educated citizens will act in the interests of society because they know that this is a way to advance their individual interests. In addition, intellectual development increases the capacity for obtaining the knowledge needed to achieve optimal outcomes. Thus, for Mill, intellectual development and social improvement go hand in hand in the attempt to produce the greatest happiness for the greatest number of people. Knowledge emerges as a central utility for achieving an ethically constituted society within the utilitarian framework.

The establishment of the USDA/land-grant system is consistent with a utilitarian policy for advancing the general welfare through the acquisition and distribution of knowledge. Although it is not clear how directly the utilitarian theorists influenced the legislation that created the system, it is apparent that the mission of these institutions are amenable to analysis within a utilitarian framework. In the mid-1800s (and well into the 20th century), the population of the U.S. was largely agrarian, and any policy or institution that advanced the well-being of the majority of farm families could be construed as consistent with the utilitarian maxim. Some of the debate that influenced the final form of the Hatch Act was explicitly utilitarian: Indeed, the only major constraint that was imposed on the system was a requirement to make research relevant to farmers (Schweikhardt & Bonnen, 1986). The high levels of applied research performed by the system constitute at least an attempt to achieve the utilitarian ideals embedded in the system's mission, and to the extent that the system actually did increase the well-being of a majority of American farmers in its early decades, it actually fulfilled those ideals.

Thus, there is a dual reason why increasing agricultural productivity through research is an application of utilitarian principles. First, if increasing productivity just means getting more of a valued good from the same quantity of land, labor, and raw materials, setting productivity as the goal is a fairly direct translation of the utilitarian maxim into production practice. Second, the emphasis on education and research would be expected to have a positive long-term effect upon society's capacity to produce benefits in the future. Although critics of utilitarianism (and many utilitarians themselves) could find areas to dispute in these broad claims, they are undoubtedly plausible when viewed in light of utilitarian principles. As such, it would have been reasonable for agricultural scientists, research administrators, and public officials to proceed on the basis of shared belief in the validity and justifiability of increasing agricultural productivity through agricultural research.

To the extent, then, that agriculture in general, and the USDA/land-grant system in particular, has emphasized increasing agricultural productivity and reducing agricultural production costs, a utilitarian ethical framework has been implicit. Although few administrators or scientists have probably ever explicitly thought of themselves as utilitarians, it is clear that they believe they are morally justified in pursuing productivity-oriented research to increase or optimize the social benefits derived from agriculture. The utilitarian maxim merely shows what they already knew in an implicit or common-sense fashion.

Research Entrepreneurship and Utilitarianism

So far, the discussion assumes a link between research and agricultural practice that needs to be more explicit. One interpretation of the link is to assume that research has increased productivity because it was directly applicable to the production problems of farmers. But, as Suppe (1987) has shown, direct applicability is rarely achieved in fact. Direct applicability at the farm level requires a close match between the soil, climate, managerial, and economic conditions of the laboratory and the farm itself. On the contrary, however, farmers are unlikely to be able to duplicate exactly the conditions within the laboratory. Rather, farmers select and adopt research results much as they would select and adopt indigenous knowledge handed down from neighbors and relatives: they adopt technologies and practices only when they believe they will assist in achieving their goals. Science, however, purports to advance knowledge more systematically than farmer trial and error. Thus, something other than direct applicability of results must link researchers and adopters. Here we are not concerned with the activities of extension agents or commercial suppliers. Rather, we are concerned with the strategy that makes innovative knowledge available to extension services and commercial firms in the first place. We suggest that this mechanism is research entrepreneurship, a process of unrestricted inquiry that allows the laboratory to choose its own goals without interference, thus determining the value of knowledge by a process akin to the market's selection of innovative firms.

Busch and Lacy (1983) use the term "research entrepreneurship" to describe scientists who successfully link their individual disciplinary research interests to broader goals by matching individual interests to those expressed in the agenda of various funding agencies. This concept aptly describes how scientists gain support for their projects. However, it does not capture the implicit values expressed in the priority of personal interest and importance to society as the two most important criteria for problem choice. Nor does it capture the sense in which advancing knowledge is a utility with a value that derives from its potential to optimize future outcomes. Thus, another form of research entrepreneurship may be defined as a research strategy that obtains ethical value from the assumption that if scientists motivated by public service interests are allowed to use their ingenuity freely, they will generate the knowledge necessary to advance the public interest.

This form of research entrepreneurship is an approximation of what Mill referred to as liberty of thought and discussion, which is closely linked to the utilitarian maxim in his essay, *On Liberty* (1859). According to Mill, (1859:18), "Complete liberty of contradicting and disproving our opinion is the very condition which justifies us in assuming its truth for purposes of action." Mill suggests (1859:32) that even suppression of false views has an insidious effect. Less-than-free debate can inhibit intellectual progress because a "fear of heresy" stifles the pursuit of independent thought:

> Truth...[is] a question of the reconciling and combining of opposites..., and it has to be made by the rough process of a struggle between combatants fighting under hostile banners....if either of...two opinions has a better claim than the other, not merely to be tolerated, but to be encouraged and countenanced, it is the one which happens...to be in a minority.... (Mill, 1859:46)

From Mill's discussion of freedom of inquiry, innovation can never take root (or do so only with the greatest difficulty) if new, unorthodox ideas cannot be put forth for discussion. Furthermore, Mill believes (1859:61) the ongoing consideration of new ideas—even false ones—is necessary to maintain creativity: "There is only too great a tendency in the best beliefs and practices to degenerate into the mechanical" unless they are constantly tested against alternatives (p. 61).

Although Mill's essay stresses the linkages between liberty of thought and the criteria by which we may be justified in using knowledge claims as a basis for action, the most relevant point is Mill's emphasis on the idea that people vary in their experiences and circumstances:

> It is for [each person] to find out what part of recorded experience is properly applicable to his own circumstances and character....[In] the first place, [others'] experience may be too narrow, or they may not have interpreted it rightly. Secondly, their interpretation may be correct, but unsuitable to him. Customs are made for customary circumstances and characters; and his circumstances or his character may be uncustomary (Mill, 1859:55).

Thus, the scientists who produce knowledge may not be well-positioned to determine the value of any given body of knowledge in any particular application: "with respect to his own circumstances...the most ordinary man or woman has means of knowledge immeasurably surpassing those that can be possessed by anyone" whose circumstances are different (Mill, 1859:74).

Concepts similar to Mill's were important in the creation of the research system. Schweikhardt and Bonnen (1986:14) suggest that the structure of the experiment station system is unique because it is "a geographically and administratively decentralized system capable of addressing the research needs of farmers facing many different ecological conditions." This decentralization has contributed to productivity growth in U.S. agriculture (Evenson et al., 1979; Ruttan, 1987). In the Hatch Act debate, amendments were introduced to give the USDA at least some control over research. These amendments were opposed by strict constructionist politicians on the grounds that they would unconstitutionally extend the authority of the federal government to the detriment of individual freedom. Strict constructionists succeeded in quashing the amendments, and in giving the states power over matters not specifically allocated to the federal government by the U.S. Constitution. Senator John Ingalls of Kansas, a strict constructionist, opposed the amendments because independence of the experiment stations from centralized control was necessary to ensure a "collision and contest between opposing views" that would produce "the greatest good for the greatest number" (quoted in Schweikhardt & Bonnen, 1986:17). Schweikhardt and Bonnen suggest that the values of research entrepreneurship and utilitarianism continued to shape the system in the years following its creation. Not only was political intervention in the stations' research activities discouraged, attempts were made to free researchers from administration, teaching, and extension duties in order to avoid diluting their research efforts. The utilitarian orientation of many scientists and administrators sustained a high level of applied research that often was instrumental in obtaining political support from farmers for research.

Decentralization of the system had several effects. The geographic dispersion of research stations among the states led to research in a diverse set of ecosystems to benefit farmers with a wide variety of geographically specific problems. Administrative decentralization provided local control over research which (together with relatively close relationships between basic and applied researchers) facilitated responsiveness to local problems, while simultaneously providing research coordination at the national level through the USDA. However, Schweikhardt and Bonnen note (1986:18) that much of the system's success stems from an historical accident: Decentralization occurred because of a compromise between scientific, ethical, and political values, not because of any notion of the requirements of a successful national research system. Thus, part of the basic structure of the laboratory resulted historically from an explicit linkage between the utilitarian maxim and research entrepreneurship as a means to achieve optimal outcomes.

As criticism of the system has increased, so have demands for greater oversight within the system (Schweikhardt & Bonnen, 1986; Ruttan 1983, 1987, this volume; MacKenzie, this volume). Theodore Schultz (1984) laments the burdens that increased accountability have placed on agricultural research. Schultz views agricultural research as an economic activity with the principal product of information, the value of which is its usefulness as a form of capital for economic growth. He claims (1984:344) that the "fundamental dynamic agent of long-term economic growth is the research sector of the economy," and that if research is to perform this function, it must avoid "static, unchanging routine work on the part of scientists." Schultz believes that research entrepreneurship is as essential as scientists, facilities, and funds for us to meet future needs. He suggests that although some administrators may be research entrepreneurs, most entrepreneurship comes from scientists at the leading edge of research because their positions uniquely qualify them to assess the potential for advancements in knowledge. Schultz notes (1984:345) that as research administration becomes increasingly centralized, identifying and rewarding entrepreneurship becomes increasingly difficult: "a highly organized research institution firmly controlled by an administrator...is the death of creative research. No research director in Washington or Santiago can know the array of research options that the state of scientific knowledge and its frontier afford." Thus, Schultz believes that a primary challenge facing agricultural research is to develop institutions that foster research entrepreneurship at the bench scientist level.

This viewpoint echoes Mill's opinions in several ways. Even though it stems partially from historical accident, the success of geographically decentralized research reflects Mill's belief that knowledge must ultimately be applied according to the circumstances of the agents involved. Administrative decentralization embodies concepts that are important not only in Schultz's view of research entrepreneurship, but also in Vernon Ruttan's view of moral responsibility (Ruttan, this volume; Ellis, this volume). On the one hand, according to Ruttan, research administrators must be active participants in an open political dialogue that establishes research agendas and funding levels consistent with social goals. This recognizes individuals in society as being best placed to determine their needs, and administrators as being best placed to communicate the economic, social, and other utilities that can be expected from agricultural research. It also recognizes the need to ensure that the research agenda reflects a political will derived from an open, well-informed debate that expresses not only what society needs, but also the costs society is willing to bear.

On the other hand, within the constraints of meeting social goals, administrators should foster research entrepreneurship to help ensure that a maximum range of scientific responses are available to meet social goals. In fostering scientific freedom, according to Ruttan and Schultz, administrators avoid impairing innovation by deferring problem choice to scientists, who are the individuals best positioned in the various disciplines to determine what kinds of advancements are possible with respect to meeting goals. This reflects Mill's belief that social needs can be met best in the long-run by freedom of inquiry and action. Moreover, Ruttan's case (1987, this volume) for a light burden of moral responsibility on scientists and administrators is similar to Mill's position with respect to self-censorship in inquiry. If moral responsibility for the results of research leads scientists to engage in mechanical, uncreative research, then the source of innovations for future needs may dry up, leading to decreased utility, and, therefore, morally undesirable consequences.

Problems with Research Entrepreneurship and Utilitarianism

The rankings of criteria for problem choice in Busch and Lacy's study are consistent with our claim that research entrepreneurship and utilitarianism are major value orientations in the laboratory. The high rankings of personal interest, scientific curiosity, and assessment of client needs by the scientist (first, fourth, and seventh of 21, respectively) suggest that scientists believe they are determining the direction their research takes. This seems especially apparent in the relatively low rankings of institutional priorities, client demands, and feedback from personnel (eleventh, thirteenth, and twentieth, respectively). The high rankings of importance to society, client needs, and development of new methods, materials, and devices (second, seventh, and fifth, respectively) suggest that scientists believe they are working toward the social good. If, in fact, research entrepreneurship and utilitarianism form the ethical foundation of problem choice, however, we must note that the foundation itself is subject to criticism on several fronts.

Stan Dundon (1983, 1986) challenges the philosophy of research entrepreneurship. Two levels of value commitment inhibit innovation in agricultural research: commitments to professional paradigms and to extant ethical standards. He believes that these commitments converge in a form of "moral orthodoxy" that is challenged by innovation. This position is based on interviews with innovative researchers which reveal, says Dundon (1986:41) the following:

> a severe tension between innovators and their peers. This tension is due to a belief by the peers that the innovators are not 'good' innovators, i.e., what they do does not fit the mold of expected innovation....The innovators themselves suspected that the tension and discomfort that their work created was due to the fact that their innovation involved questioning some fundamental commitments....(1986, p. 41).

Dundon believes that most agricultural scientists are committed to a view of research which includes a sincere concern with ethics that usually involves service to particular client groups. However, within well-defined disciplines, an innovator may be limited because the introduction of new practices challenges other researchers' professional competence and ethical commitments. Innovation implicitly questions ethical commitments to existing client groups by identifying new or under-served constituencies. Professional competence is questioned because, Dundon suggests (1986:41), the research profession includes the notion that "'true' innovations will be new tools and new procedures that fit the professional mode of practice." Researchers committed to a narrowly defined disciplinary focus may be biased against innovation because evaluation of it by other members of the discipline may be partly or largely outside their range of professional training and experience. Thus, Dundon suggests that commitments by researchers may be a significant barrier to innovation, echoing Busch and Lacy's observations that disciplinary values are pervasive, and that the disciplines have minimal contact with each other, and echoing Mill's claim that originality is hard to recognize as such.

Dundon's work on research choice provides a reason for embracing research entrepreneurship advisedly. Research entrepreneurship allegedly serves the greater good because researchers are free to pursue their scientific activity without restraint or direction. If their research results are selected by members of a given client group, it is because the adopters have interests of their own that they believe will be advanced by new knowledge or technologies. In one sense, the philosophy of research entrepreneurship denies the importance of linking the scientifically goal-oriented activity that takes place within a laboratory to any broader goals at all. If scientists produce enough knowledge, some of it will be useful to somebody. Scientists will produce more knowledge if they are left alone, so we (e.g., society) maximize our chances of gleaning something useful from scientific laboratories if we resist the temptation to direct them toward one goal or another. In other words, because only scientists are well placed to determine how and in what directions the frontiers of knowledge can be advanced, if we allow them to do what they do best, knowledge will automatically become available for application toward valued social goals.

The problem that Dundon notes with this model is that values are already built into scientific activity. We have characterized this problem in terms of the laboratory: values are built into agricultural laboratories in the form of equipment and personnel choices, not to mention disciplinary training and scientific paradigms. If Dundon is correct, the laboratory character of agricultural research policy may subvert the strategy of research entrepreneurship. Research choice is already highly structured by disciplinary and organizational values. Within any given laboratory with a given disciplinary orientation, then, the most that research entrepreneurship can guarantee is that laboratory personnel will produce the maximum amount of knowledge that is consistent with the physical, theoretical, and goal constraints built into the laboratory. Because any combination of physical, theoretical, and goal constraints can serve only a limited range of values, it would be a remarkable coincidence if research entrepreneurship in laboratories that have been producing knowledge primarily valuable to the "large farm" could suddenly begin producing knowledge valuable on altogether different terms. Thus, if research entrepreneurship is to be the strategy that guides problem choice, the conceptual components of the laboratory may need to be drastically reequipped. Indeed, the critics' call to address alternative research streams is equivalent to a call to reestablish a broadly based research entrepreneurship in the laboratory. As such, this also constitutes a call to comply with the values invoked to justify the original structure of the system.

Moreover, despite their initial plausibility, utilitarian decision procedures have historically been plagued by a number of philosophical problems. Some of these problems, though familiar to philosophers as general questions for utilitarian moral theory, raise particularly important concerns for agriculture. Of these general concerns, an empirical problem in anticipating consequences, and a moral problem relating to the problem of making trade-offs are particularly relevant.

One general objection to optimizing decision procedures is simply that obtaining complete and reliable information on all the relevant consequences of a policy decision is impossible in practice. Some factors are inevitably omitted, and, when these factors affect human health and safety, the economic well-being of minority groups, or environmental quality, the entire moral calculation of relative benefits can be drastically modified. Costs or harmful effects that are simply left out of a utilitarian calculation are called *externalities*. Such costs are sometimes left out either because the decision-maker does not have a reliable way to measure or compare them, or because the decision-maker does not have to bear them. When UC researchers decided to develop the tomato harvester, they could have reliably predicted that some growers and workers would experience adjustment costs as they found (or failed to find) new forms of employment. In comparison to the estimates of research expenditures, these costs would have been difficult to measure; if we count the emotional stress suffered by dislocated growers and their families, calculating these harmful effects would be impossible. Furthermore, these costs were not borne by UC, but by the growers and workers themselves. UC decision-makers, therefore, had a double reason not to count them.

From an ethical perspective, however, these costs must be considered for the decision to be justified by the utilitarian maxim. Decision-makers must make stringent efforts to reflect all such "externalities" in any estimate of social benefit or harm, or they cannot truly be said to have optimized outcomes. The "neglected dimensions" noted by the contributors to Kenneth Dahlberg's 1986 book on agricultural research are examples of such externalities—harms or costs that might alter the evaluation of productivity-oriented research streams, if they were properly "internalized." Among the topics cited by Dahlberg authors are long-term energy costs of high input production techniques (Pimentel & Pimentel, 1986), social impacts of the changing farm/size distribution upon rural communities, emotional stress suffered by financially troubled farm families (Heffernan, 1986), and health effects, both for agricultural workers and for consumers of agricultural products (Coye, 1986).

Decision-makers with a single-minded focus on increasing crop yields may neglect adverse effects relating to nutritional quality of food, impact upon small farms, or deteriorating soil fertility and water quality. The mere fact that one lacks data, or will not have to personally pay these costs, however, is no rationale for ignoring them. Although the problem of externalities is administratively difficult to resolve, it is philosophically simple, in that the way these uncounted harms must be weighed in comparison to counted benefits is unambiguous. It is a problem of failing to count all the relevant outcomes at issue here, rather than not knowing how to count them. Mentioning and accounting for these costs is not a *revision* of agriculture's utilitarian approach to ethics; it is merely a *completion* of the benefit-cost calculation. As such, the critics' call for a consideration of nonproductivity issues in agricultural research is equivalent to a call for scientists to consider a more full and realistic set of costs and benefits. As such, this constitutes a call to comply with the utilitarian values that were invoked to justify the existence of the system.

Many suggest that complying more fully with utilitarian values is not enough; the problem, they might claim, is utilitarianism itself. Emphasis upon pure productivity levels (or other measures of utility) neglects distributional questions, such as the utilitarian maxim itself. That is, utilitarian evaluation of agricultural research may appear to place total emphasis on the aggregate bargain of costs and benefits, and no emphasis on how the costs and benefits are distributed. Clearly, the use of public funds to benefit the rich at the expense of the poor violates the basic principles of fairness and democracy (DeWalt, this volume). However, merely completing the benefit-cost calculation will not always rectify such a situation. If the completed calculation benefits the rich without benefiting the poor, the outcome is still moral in utilitarianism.

As such, a more pointed statement of the problem with utilitarian research planning might claim that agricultural research has increasingly benefitted large farmers and middle class food consumers, two groups that are fairly well off, while harming small producers and farm labor, two groups that are less well off. Such a statement would claim that the fact that the research produces more benefits than costs is not relevant because the ethical issue is one of fairness—the use of public resources to help the rich, rather than the poor. Furthermore, much stronger moral complaints exist against utilitarianism. Agricultural research could culminate in activities that violate the rights of individuals, or cause them to lose control over their lives in philosophically fundamental ways. For example, a cost-benefit argument may attempt to evaluate a new chemical by comparing the moral cost to the health of a few agricultural laborers with the pennies saved by millions of food consumers. Although the benefit for an individual consumer may be trivial, when multiplied several million times, it might be considered to outweigh significant harmful effects on a few workers. A utilitarian approach would not rule out the possibility that significant benefits from the research might outweigh even these extreme forms of harm. These latter criticisms are a basis for rejecting utilitarianism altogether.

Ethics in Agricultural Research: Beyond the Large Farm

A question that emerged repeatedly during the gestation of this book is the following: If agricultural research abandons its utilitarian foundation, what ethical foundation will it adopt as a replacement? Utilitarianism has always been an object of criticism from other ethical perspectives. However, none of the other traditions has been able to deliver a decisive defeat, and utilitarianism remains one of the enduring ethical traditions of the modern age. MacIntyre (1989) suggests that an ethical tradition may retain its appeal, at least partially, because it is capable of elucidating and addressing the kinds of problems its adherents typically face. Although none of us is particularly committed to utilitarianism as an ethical theory, we nevertheless maintain that it may be the most appropriate framework in which to address morally relevant problems of agricultural research policy. Thus, a less drastic approach than abandoning utilitarianism would be to recognize certain constraints upon utilitarian decision-making, or, alternatively, to recognize that certain goals should be sought even when the means for achieving them are not cost-effective, at least when only monetized values are considered.

Utilitarianism is compatible with a broad range of alternative values, whether it accepts them as compensating costs and benefits, incorporates them as previously ignored externalities, or regards them as constraints on allowable outcomes. Framing the discussion of ethical aspects of agricultural research policy in terms of utilitarian values makes the discussion intelligible to scientists and administrators in terms of the values that are already established in the laboratory. At least some of the solutions may be compatible with the nonutilitarian foundations that initiated the criticisms in the first place. In other words, a tremendous amount of latitude exists within the utilitarian framework of agricultural research to allow scientists and administrators to expand the range of values served by research without paralyzing the system, while its members wait for philosophers to formulate the perfect ethical theory.

If our account so far is more or less correct, then the traditional emphasis of agricultural research on productivity is an expression of the utilitarian and research entrepreneurship values that were legislated into the system. These values appear to remain central to the research enterprise. This is not to say that all agricultural scientists are utilitarians or research entrepreneurs; it suggests only that their research is influenced (or at least rationalized) by the incorporation of these values into the agricultural laboratory itself. To the extent that scientists have employed research entrepreneurship in an attempt to comply with the utilitarian maxim, criticisms of the historical focus on productivity are not necessarily criticisms of utilitarian values, but rather, criticisms of the traditional measure of utility. Therefore, it follows that many ethical criticisms are directed neither at the scientific foundation of agricultural research nor at the moral character of agricultural scientists.

In summary, a burden is imposed upon research planning based on the laboratory analysis given above. The laboratory analysis of agricultural research policy shows why a change in the broad goals of agricultural research may be more difficult to achieve than it seems. Because research facilities are structured by their equipment, personnel, and scientific goals to pursue certain research streams (and not others), they cannot easily or spontaneously change the research products they produce. For example, an assembly line built to produce bombers cannot convert effortlessly to the production of automobiles on the day peace is declared. Converting the productivity laboratory to research on farm preservation is probably more difficult than converting the assembly line. The philosophy of research entrepreneurship, scientific curiosity, and the scientist's intellectual interests are parts of the equipment of a research laboratory for which there are no analogs on the assembly line. If the scientific work underway in an agricultural laboratory truly supports the goal of increasing large farm productivity, it may be because large farms can use the general kind of knowledge that can be produced by that laboratory, and not simply because scientists have adopted a utilitarian rationale for increased productivity as the broader goal that justifies their activity. An obvious response is that agricultural scientists should at least diversify the broad social goals that their research is expected to advance. It is in this sense that the essays contained in this book are intended to take us "beyond the large farm."

Thus, an additional option should be considered in research planning—one that maintains the traditional commitment to serving the public good, but aims to rethink the recent definition of that commitment in productivity-enhancing terms. The configuration of our current laboratories in agriculture places a heavy burden of proof on scientists who opt to pursue alternative goals. However, scientists themselves can rethink the values that shape their daily practice, as well as the research that serves those values. Perhaps without even realizing it, they have the power to change the values served by agricultural research. What is more likely is that laboratories will continue to do what they always have, but will call it by a new name (sustainable agriculture, perhaps), or researchers will engage in philosophical and political defense of the values by which they have justified their activities in the past. But we have already seen enough of that, so we turn now to the more hopeful (if also less likely) alternative.

Essays Toward a New Vision of Agricultural Research Policy

The essays that follow in this book attempt to begin the process of establishing a new vision of agricultural research policy. The authors differ regarding how far from the traditional course a new ethic of research planning must steer. But they are united by a belief that change is necessary and a hope that it is possible.

In his chapter, "Agroethics and Agricultural Research," plant pathologist David MacKenzie claims that the shared cultural environment of the farm family served as the basis for an implicit ethic that guided agricultural research for nearly 100 years. Specific standards of fairness, responsibility, and duty to constituency did not need to be spelled out because agricultural researchers and their farm constituents shared an understanding of what was expected and allowed. He suggests that this implicit knowledge is crumbling because (1) farm families now make up less than 2 percent of the population, and (2) publicly funded agricultural researchers are under increasing pressure to capture a portion of the commercial value of their research. MacKenzie perceives that an acute need exists for researchers to think explicitly and carefully about the ethical goals and responsibilities of their work. He doubts the ability of the market and the political process to implement appropriate standards and makes a plea for minimal regulatory oversight of agricultural research.

Agronomist Vernon Cardwell proposes new directions for agricultural research in his chapter, "Balancing Moral Imperatives Through Rural Development." He views agricultural problems as problems of rural development that are grounded in the production and social concerns of families. As such, research problems can be identified and priorities set through an interactive and iterative systems approach. Research teams interact with farmers to discover how production problems and family/community concerns are interrelated, and then propose solutions compatible with production and social needs. The process is iterative because proposed solutions may be modified in response to evaluations by farm families and/or changing rural conditions. Cardwell believes that this approach is essential to bringing moral imperatives into balance with free enterprise.

Jeffrey Burkhardt, a philosopher working within the Institute for Food and Agricultural Science at the University of Florida, reviews some of the arguments that have been used to suggest that agricultural research has failed its historical mission to serve American family farmers. In "The Value Measure in Public Agricultural Research," Burkhardt agrees that the USDA/land-grant system has tended to favor large farmers (as well as agribusiness input suppliers, chemical companies, and grain traders), but questions the agrarian critics' judgment that this makes scientists and administrators morally blameworthy. He concludes by arguing that agricultural research planning must be influenced by a broader cross section of the American public. Food consumers, small farmers, and the working poor must all be served by agricultural research. Achieving this goal requires explicit goal-setting with an eye toward a more philosophically sophisticated concept of the public good.

Agricultural economist Vernon Ruttan emphasizes two points about the importance of ethics in his essay "Moral Responsibility in Agricultural Research." First, values that determine the relative importance of cost-efficiency, land-saving and labor-saving technologies, and broader social goals form what he calls the cultural endowment. The cultural endowment determines which agricultural technologies are perceived as viable by researchers and the public alike. As such, moral values have an influential role in the development and diffusion of agricultural technologies. Second, he makes a plea for more explicit research on a series of value issues, including research on how notions of fairness might be applied to evaluating technology and research. He rejects the idea that assessments of expected utility should be used as a basis for research decision-making; rather, utility assessments should serve as the basis for political debate that establishes the research agenda. Furthermore, he argues that individual scientists should bear a light burden of moral responsibility for the results of their research.

Entomologist Miguel Altieri offers a comprehensive discussion of sustainability and agricultural research in his essay "An Agroecological Analysis of the Environmental Degradation Resulting from the Structure of Agriculture." Altieri discusses how peer review standards in science affect research choice. He suggests that meaningful research on sustainable agriculture requires the simultaneous adjustment of many variables, including mixing crops, altering production techniques, applying indigenous knowledge, and even engaging in political action that affects such areas as market development and access to credit. Altieri claims that applying a standard criteria of research design to such a complex system of variables is impossible: When one obtains success, assigning causality to any specific component of the system is sometimes difficult. Peer review standards that require demonstration of causality or statistical significance lead agricultural scientists to select research topics for which most variables can be subjected to laboratory control. Extension of this research often includes extending laboratory control to the level of commercial production in the form of industrialized agriculture. As such, values for assessing the quality of scientific research have an unintended effect on commercial agriculture.

In "Making the Land Ethic Operational: Toward an Integrated Theory of Environmental Management," philosopher Bryan Norton discusses the difference between "conservation" values and "preservation" values as they apply to land use planning and, indirectly, to agricultural research. He suggests that much of the controversy between environmental activists and agricultural researchers has a semantic basis that can be overcome to reveal substantial areas of common agreement. Norton further suggests that hierarchy theory in contemporary ecology can be invested with operational significance for agricultural research planning: Ecologically determined hierarchies provide a framework which research planning must accommodate. He argues that the increased cost efficiencies in one level of a hierarchy can never outweigh the impacts of any sort at a higher level of that hierarchy.

In "Ethics of U.S. Agricultural Research in International Development," agricultural engineer Merle Esmay claims that development plans often fail because they are based on technology packages inappropriate to Third World conditions. The tendency to introduce such packages results from the lack of a broad vision. This problem is especially acute for Third World Ph.D. students for two reasons. First, their First World education mirrors the basic research emphasis and high technology facilities of Western agricultural research. Consequently, they may not be prepared for the applied research focus and low technology facilities they face on repatriation. Second, the specialized research background they obtain does not prepare them for the management or government roles they often assume soon after repatriation. Esmay believes that we must recognize that the needs of Third World students are significantly different from Western students, and develop programs specifically tailored to students from developing nations.

In "The Cultural Ecologist Concept of Justice," anthropologist Billie DeWalt suggests that the value issues raised by agricultural research for developing nations are often disregarded. He argues that this trend has existed for many decades, and that criticisms of Green Revolution strategies have at best been only partially accommodated in agricultural research. He claims that research oriented around the production of greater quantities of food by fewer people makes little sense when a billion people already have nothing productive to do to gain access to increased food supplies. DeWalt suggests that the client group for publicly funded agricultural research must be poor people. He further suggests that U.S. universities should create multidisciplinary research organizations devoted to agricultural research for developing nations.

The next three chapters are bibliographic essays. Each attempts to review and summarize the publications written by agriculture's critics during the last two decades.

Nancy Shankle's "Agricultural Research Policy and the Family Farm" reviews the literature on agrarianism as a theme in American life. She summarizes arguments from a number of agrarian authors who have defended the view that preservation of small family farms should be a goal of agricultural research.

Paul Thompson, Gary Varner, and Deborah Tolman's "Environmental Goals in Agricultural Science" is a bibliographic essay of the value issues in environmental ethics as they relate to application in agricultural research. This chapter begins with a brief discussion of environmental impact and animal welfare as externalities of the agricultural production process. The main emphasis is upon authors who reject the philosophical framework that is implied when impacts upon animals or upon nature are understood as "costs" of agricultural production. The essay summarizes the environmental ethics issues that confront agricultural research and introduces the reader to a wide range of sources.

"U.S. Agricultural Research Policy and International Distributive Justice" by Lain Ellis is a bibliographic essay intended to provide an introduction to the value issues in the relationship between research policy in the U.S. and international justice. Ellis focuses on identifying the value-related concepts that underlie alternative views of agricultural research for Third World countries. Ellis also provides a brief introduction to nonutilitarian concepts of justice as they relate to research policy.

The book concludes with "Beyond the Large Farm," a chapter that reassesses the problem of external forces on agricultural research by agricultural engineer Bill Stout and philosopher Paul Thompson. Stout and Thompson explore two points of view that can be taken with respect to external critics of agriculture. One, the view from the bench, represents the perspective of a working agricultural scientist. The other, the view from nowhere, represents the vantage point of a cultural observer who has no stake in the agricultural system. Stout and Thompson conclude by calling for generalist research and education which can bring insights obtained from these specialist viewpoints to bear upon agricultural research decisions.

Notes

1. In particular, one may question the *causal* allegation that research affects changes in the size distribution of farms (see Martin & Olmstead, 1985), or the *normative* evaluation that it is large farmers who are the primary beneficiaries, as opposed, for example, to food consumers (Madden & Thompson, 1987).

1

Agroethics and Agricultural Research

David R. MacKenzie

Introduction

A tension is rising in the agricultural research community that needs to be addressed. This tension is a result of new expectations in the way public sector research discoveries are distributed. Many public institutions are now viewing research discoveries as patentable "intellectual property" with the potential for earning royalties to help them pay their bills (Wolfsy, 1986). This issue has become particularly important to U.S. colleges of agriculture, many of whom have barely kept pace with inflation since the 1950s.

Some recent developments are contributing to the feasibility of this approach. Exciting advances in molecular biology promise dramatic genetic improvements in crops and livestock. In addition, new interpretations of the U.S. patent law permit the patenting of life forms (Buttel & Belsky, 1987; Evenson, 1983; Evenson & Putnam, 1987). So why does tension exist in the agricultural research community about these proposed changes? Because they challenge the traditional ethical values of the public sector agricultural research scientists.

The purpose of this chapter is to explore the sources of these ethical values in the context of public sector agricultural service and to offer interpretations and suggested solutions for these contemporary dilemmas that have wedged progress against traditions in agricultural research.

For illustration, agricultural biotechnology research will be used to convey some of the broad issues relevant to this topic. Even though biotechnology research in and of itself is not the cause of the current tensions—it is one of many new technologies that have opened both the floodgates of institutional opportunity and the deep-rooted ethical concerns of the agricultural science community.

The Concept of Ethical Heritage

Ethical values, or the patterns of choices that we make for appropriate action, are transferred from generation to generation. We learn these standards from the people with which we associate. Hence, they are "socially heritable." One's family upbringing is a strong source of such standards, but so is society-at-large. Ethical values also evolve. Codes of behavior change with time and in response to the requirements of new circumstances. The ethical values that would have been defended by our forefathers are vastly different from those defended today.

Ethical choices are not absolute choices, good or bad, but they are nominative behavior patterns consistent within social groups. One set of values is not more ethical than another, and each set must be viewed in the proper context. For example, the U.S. farm community is generally highly regarded for its high ethical standards. But, some exceptions exist, such as environmental issues. To some, it is curious that contemporary agriculture is not a leader in the environmental movement. After all, the agricultural industry is absolutely dependent upon the preservation of its natural resources. But the farmer/producer accepts the utilitarian ethical need to both increase production and ensure a steady supply of food and fiber (Thompson, 1988b). This is accomplished by using specific technologies such as plant fertilizers and pesticides, which conflict with the values of the environmental movement. Neither can be judged right or wrong—only different in their priorities.

Ethical values also differ among institutions. For example, within the public sector, the assignment of authorship at universities on all types of documents follows predictable patterns. However, the individuals credited for writing government service documents are commonly the highest level administrator with relevant responsibility, and little or no reference is given the ghostwriters of the documents. Although this practice is commonly accepted in government service, it is considered unethical to the academician. Thus, as individuals with different backgrounds move between social institutions, some of the ethical standards seem to change. These new standards must be accepted if one wishes to "fit in" (Pollock-Byrne, 1984).

An enormously important condition is clear, however. Ethical systems must not change too easily, for if they do, they cease to function as a norm and cannot guide our actions in a meaningful sense. If ethical standards change too rapidly, they fail to link our lives to the traditions that have given meaning to human activity throughout several generations. Thus, the thesis of this chapter is that the U.S. agricultural research community has a heritage of ethical standards with elements that are unique to agriculture and traceable to the value systems of the American farm family.

Agricultural Heritage

Dr. Donald F. Jones, plant geneticist with the Connecticut Agricultural Experiment Station (1915-1960) and a member of the U.S. National Academy of Sciences is credited with two major genetic discoveries. Becker (1976) recounts these contributions. Jones' first discovery (at age 25) was the development of double cross hybrid maize which proved to be an enormous boost for maize yield worldwide. His second discovery (at age 60), with Dr. Paul Mangelsdorf of Harvard University, was the practical application of cytoplasmic male sterility (CMS) for the commercial production of maize hybrid seed without the expense of hand detasseling. Together Jones' two discoveries revolutionized U.S. and worldwide maize production. As a result, Jones and Mangelsdorf, through their respective institutions, were awarded U.S. Patent 2,753,663 on July 10, 1956, for their invention of CMS-produced hybrid maize seed.

During the annual business meeting of the American Society of Agronomy in Cincinnati, Ohio, in mid-November, 1956, Dr. Jones was called to the podium. Well-known for his contributions to maize genetics, he might have expected this "surprise" to be yet another professional society citation for his scientific contribution, but it was not. It was a reprimand by his peers for patenting the CMS technology. The resolution read as follows:

Special Resolution

Cooperation in agricultural research has become a motivating force in technical progress. Plant breeders have been among the leaders in such cooperation. The entire hybrid corn industry is a monumental tribute to that cooperation.

A severe blow has been dealt to such cooperative endeavor by the application for Patent No. 2,753,663, granted July 10, 1956. The method outlined in this patent now issued, and in an application No. 47912 now pending, was the direct outgrowth of research contributions by many individuals and through free exchange of information and germ plasm among corn breeders and plant scientists.

The capitalization on such information by any one individual or individuals thus becomes a breach of faith in this principle of interchange

of information and material, and seriously jeopardizes the future contin-
uation of such cooperative endeavor.

Now, therefore, be it resolved:

That this expression of viewpoint by the American Society of Agronomy become a permanent record in the minutes of its meeting held in
Cincinnati, Ohio, on November 12 to 15, 1956, and that copies be trans-
mitted to:

Donald F. Jones

Paul C. Mangelsdorf

Research Corporation of New York

The Association of Directors of the Land-Grant Colleges and
   Universities

Harvard University

Director, Connecticut Agricultural Experiment Station

The American Seed Trade Association

The American Society for Horticultural Science.

This resolution was approved also by the Crop Science Society of
   America (Anonymous, 1957).

Jones (the agricultural scientist) had violated the ethics of the professional society membership for patenting his invention of CMS-produced hybrid maize and he was publicly scolded. Many say that Jones
subsequently donated the money earned from the patent's royalties to
charity.

A comparable invention by a university-employed engineer likely
would not have met a similar fate. University poets publish copy-
righted books and gather royalties; chemists claim patents; football
coaches charge speaking fees; professors of business collect consulting
fees; and administrators accept honorariums. But for public service in
agriculture, any form of renumeration beyond salary is considered
professionally unethical, except for extraordinary circumstances. Harvard Professor Mangelsdorf, who shared in the patent with Jones, did
not experience the same humiliation from his peers. Harvard University
is a non-agricultural institution with distinctly different ethical standards.
Where did agricultural research get this tradition and where is it taking
us?

American Farm Family

The American farm family is not only a production unit but a social
unit as well. Stereotypical American farm families are hardworking,
skilled, and inventive. These characteristics are combined with strong
independence, high moral character, trustworthiness, and a sense of fair
play.

Other parallel values can be observed at the community level. American farmers are considered to be supportive of education, they have strong democratic values, and they actively support church activities. These attributes, when combined with community pride, civic duty, and neighborliness have become common in American rural life.

Holdeman (1983) writes about some of these characteristics, which he says make up the "good Samaritan." The willingness in the farm community to help others, he says, is traceable to the western migration of the nuclear family that left the extended family behind that would have otherwise helped in times of need. "To survive, the pioneer had to form mutual aid relationships with strangers. The early pioneer west of the Allegheny Mountains realized 'if I don't help you and you don't help me, then both of us will lose—the wilderness, hardships, and Indians will defeat both of us.'" Thus evolved the traditions of mutual assistance with no cash exchanging hands, as communities were interlinked not by blood, but by need.

I have termed this American farm family value system *agroethics*. The topic of "farm family values" has received significant attention by a number of authors, most of whom argue that something should be done to protect the American farm family from extinction. Although other scholars argue against such a distinct classification and question the uniqueness of agricultural ethics (Burkhardt, 1988), the public clearly recognizes such a distinction. Special treatment of American farm families in books, films, and public policies suggest that we accredit them with some special values (Madden and Thompson, 1987). What is not clear is the intent of the treatment. Is it to preserve the farm families themselves, or the value system they represent?

Many believe that a farm upbringing helps to provide ethical values for subsequent generations. This is, of course, consistent with the earlier assertions that ethics are socially heritable. Some interesting dynamics go along with this assertion as the demise of the American farm family is considered a consequence of new technology followed by harsh economic realities.

Farm Exodus

Two centuries of agroethics have strongly affected the broader range of American values, politics, and public and private life. The intergenerational transmission of these values is traceable to the American farm family. A hundred years ago, nearly half the U.S. population was directly engaged in agriculture. But over the years the percentage of Americans directly involved in agricultural production has declined to 2 percent. This "megatrend" in U.S. agriculture has been witnessed as a mass exodus from the farm, and with it, the dissemination of a value system for individuals now in careers such as teaching, athletics, science, business, law, art, trade, etc.

Many say the mass exodus from the American farm and the resulting dissemination of agroethics is the result of new, innovative technology that increased farm efficiency and thus reduced the need for labor. In other words, the American agricultural production system has been shifting to a new economic equilibrium. Although the increased efficiency of American farms has resulted in lower food costs for consumers, hardships have emerged for many farm families who were forced to give up farming and accept other circumstances.

The reservoir of farm families within our society has diminished to a very small proportion of the population. Although the public has accepted romantic notions of the American farm family and the need to preserve the institution, the wellspring is drying up. The 50:1 ratio of non-farm families to farm families has stimulated broad national concerns. This situation presents some particularly difficult problems for agricultural research.

Traditional Agricultural Research Mission

The U.S. system of public sector higher education, research, and service is unique in many respects. The land-grant universities were initiated at the height of the Civil War to provide education for the masses in agriculture and the mechanical arts. Some of these institutions still preserve that heritage in their formal institutional names which include the notation "A&M" (agricultural and mechanical). Subsequent to the creating of these institutions, research to address agricultural production problems was met through the creation of a national network of state agricultural experiment stations, mostly located at land-grant universities. The third component of the system, public service, was initiated early in this century as the cooperative extension service to provide what is now called "technology transfer" for agriculture.

Not surprisingly, many individuals forced to leave the farm during the past century found careers in this expanding network, serving as teachers, scientists, and extension agents for the agricultural sector. This turn of events no doubt provided the ethical linkage for contemporary public sector university professionals and a suitable perspective for setting the agricultural research agenda. Given the advent of the farm exodus, however, this linkage should no longer be taken for granted.

In the past, most U.S. agricultural research efforts have been a three-way partnership: (1) farmers, (2) governments (as research funding sources), and (3) public sector agricultural scientists. Through the political process, the farm community has sought resources from state and federal governments to develop technological innovation. Public funds were used to support research, commonly at the state land-grant universities. The resulting technology was then transferred (usually by the state's cooperative extension service) to the farm community free of charge. (Herein lies the ethical infraction committed by Dr. Jones and his patent on the CMS technology.) This triangular arrangement established in the public sector has operated for the past 100 years and has proven to be suitable for conventional research.

We can accurately trace the relationships of this public sector triangle, but, in recent years, the triangle has experienced heavy criticism. Many believe that technological efficiency has caused many American farmers to go out of business. Technology once considered an advantage is believed by many to be a threat to the survival of the American farm family. To appreciate this paradox, one must look at the technology adoption/benefit relationship. Early innovators are quick to adopt a new technology and increase their production (or profits) at a time when prices are relatively high. But those who are slower to adopt a new technology find themselves at a competitive disadvantage. As more and more producers adopt the new technology, unit prices begin to decline. Late adopters find that they must employ the new technologies, not to reap additional profits, but merely to remain competitive. Those who fail to adopt the technology continue to produce the same commodity volume but must sell to a market offering a lower unit price. Experience has shown that farmers who consistently resist new technologies ultimately find themselves with too little profit to justify continued operation. Thus, they go out of business and join the farm exodus.

The ultimate beneficiaries of the "technology treadmill," as it is called, are the consumers who have continued to pay lower and lower prices for abundant and good quality food. The technology treadmill has often been indicated as one of the primary causes for the demise of the American farm family (Hightower, 1973). These critics fail to note that the treadmill process is not unique to agriculture. For example, medical doctors do not receive the benefits of new medical technology; instead, the benefits of medical research are passed on to consumers, i.e., patients. Preventive medical technology means fewer patients for medical doctors, yet no one suggests that preventive medicine research has been a failure. The explanation offered for this difference in judgment is simple: Agricultural researchers have done a poor job of communicating the benefits of their work to consumers. The primary beneficiary of new agricultural technology is the non-farm sector: 98

percent of America. As critics continue to dwell on the detrimental effects of technology on American farmers, they ignore the benefits to the public-at-large. Thus, although the technology treadmill has resulted in intense competitive pressures that have reduced the number of American farmers (and hence, the wellspring of agroethics), the development of new technology has occurred in a manner consistent with the community service/public benefit ideals of agroethics.

Role of the Private Sector

Private sector research has been responsible for many innovations in agricultural technology during the last century. In particular, new chemical technologies for promoting plant and animal growth and controlling pests and diseases have emerged from private sector research. When new technologies have been marketable they too have often been developed outside the publicly funded research system.

Two points about the role of the private sector are particularly important. First, the private sector has been very effective in producing and marketing technologies if a commercial incentive has been identified. Second, the arguments of Jim Hightower (1973) notwithstanding, the public sector's association with private sector research actually has been minimal. The interaction between public and private sector agricultural scientists has mostly been in the verification of each other's discoveries.[1]

The amount of private sector support for public sector agricultural research also has been relatively small. Traditionally, improved production practices, new varieties, new breeds, and other types of technology, when developed through public sector research, have been given to the farm community free. Commercial exploitation of research discoveries has almost exclusively been confined to the private sector, in which the possibility of such exploitation is a necessary prerequisite for undertaking commercial research.

Publicly sponsored agricultural science has, in fact, been impoverished by this lack of cooperation with the private sector. No true public-private sector partnership exists. But this is changing. During recent years, a gradual shift has occurred from a conceptual triangle to a square, as agricultural technology becomes more sophisticated and private sector agribusiness grows in importance to the farmer.

Emergence of Biotechnology

In the early 1970s, rapid advances in molecular biology led to the development of technology that permitted the insertion of foreign DNA into organisms. This "genetic engineering" technology offered the promise of reconfiguring agricultural organisms to achieve improved characteristics such as improved yield and greater pest and disease resistance, as well as production of novel products, such as pharmaceuticals. Naturally, the research community ignited with interest only to stall for lack of adequate funding. The sustained costs of conducting biotechnology research is probably 5 to 10 times greater than conventional agricultural research. Start-up funds for a single new biotechnology project commonly exceed $250,000.

Coincidental with the advent of biotechnology, federal (and commonly state) funding for agricultural research stagnated. When adjusted for inflation there has been no real growth in federal support for agricultural research since the mid-1950s. To meet their new funding needs, public institutions are now looking for new partners who also recognize the enormous potential for commercialization of the products of agricultural biotechnology. Private companies are viewed by universities as new sources of support. But understandably, partnership agreements come with conditions. Today, much of the "intellectual property" of agriculture that was once given away to the farm community by public sector researchers is now viewed as patentable. New technologies are seen by public institutions as a potential source of royalties to be shared with the "partners," (i.e., the source(s) of funding, the research institution, and yes, sometimes even as payments to the principal investigator). These new "partnerships" represent a major change to the historical relationship that once existed between the farm community, the public funding sources, and the public institutions conducting the research.

The new research partnerships are not universally welcomed. Strong views in agriculture, such as the importance of public service; the traditional rights of public domain; and the concept of public trust, are challenged by the new alliances. Historically, the important ethical values of public sector agricultural research did not allow university agricultural scientists to charge fees to farmers for advice, or to accept royalties for discoveries made in publicly funded research projects. These values are now being challenged. Today, public institutional claims protect "confidential business information" as university scientists work closely with private corporations. This has raised many eyebrows and questions both on campuses and in the farm community. As one farmer and a member of a land-grant university board of supervisors said recently, "We have already paid once for the agricultural experiment station, why should we now be expected to pay royalties?" The legitimate questions are, "Who owns a patent derived from publicly funded research?" "How will tomorrow's patent technology be delivered to the farmer?" Farmers who believe that they have "paid" for experiment station research with tax dollars resent being required to purchase the rights to use experiment station-developed technology.

Public access to agricultural science is only one of the ethical issues that arises as a result of our changing agricultural research environment. As restrictions are placed on the intellectual discoveries of agricultural science, a hog farmer who may have purchased a genetically engineered hog will ask, "Who owns the piglets borne from the engineered hog?" The U.S. Congress is deliberating solutions for the anticipated problems of patented animals in agriculture. At the same time, the U.S. Patent and Trademark Office is trying to decide if specific genetic sequences— or genes, or specific genetic constructs—or chromosome segments (RFLPs), or entire genomes are patentable. And if so, which ones? Will those patents cover the offspring of transgenic animals? And will patent protection eventually extend to humans (or their genes)?

Rethinking the Ethics of Intellectual Property

This turmoil calls for reexamination of the ethical standards of public sector agricultural research. The earlier, traditional relationships between the farm community, government, and public sector research institutions can be considered a direct consequence of the agroethical value system. Many individuals (and their children) left the farm for public service jobs. These farm exodus refugees understood the expectations of farm traditions and knew the allowable choices. They understood the value system of the American farm family.

But today, these relationships are not so clear-cut. The new public and private sector partnerships are placing stress on that ethic. The role of the public service agent is changing. Some public sector agricultural researchers now accept attractive consulting fees that have evolved from their research responsibilities. Such changes are perhaps to be expected, given the observation that ethics change in response to technological, socioeconomic and cultural forces.

But there is another aspect to the current changes in the agricultural research value systems. In the old triangle, there was an information feedback mechanism that resulted from certain close relationships, particularly through the cooperative extension service, which provided direct communication between farmers and researchers. This feedback mechanism was partly political. Formula funding for the state agricultural experiment stations required scientists themselves (or through the cooperative extension service) to explain to the farmer the benefits and consequences of their research. In this way, both the negative and the positive impacts of discoveries were understood. In the emerging system for funding, will the agricultural producer be removed from this communication loop? How will priorities be set? Will all significant agricultural production problems receive research priority, or only those that have commercially viable solutions? Who will provide the non-vested interest evaluation of tomorrow's agricultural technology?

New types of partnerships will undoubtedly evolve out of the changing circumstances of public sector agricultural research. Biotechnology research makes this obvious. The national challenge will be to maintain the successful farm-government-university partnership while accommodating the new. But this will not be an easy task. The traditional understanding of "support for support" will be argued, and the ethics of the new arrangements will be questioned. However, if structured correctly, an expanded network of partnerships could mean new and critically needed resources to support public sector agricultural research.

Biotechnology Issues and Agricultural Research

We are now, as a nation, engaged in a debate about the proper choices relating to the environmental, socioeconomic, and religious consequences of biotechnology research (Farrell, 1989). Each of these three areas requires a review of traditional agroethics as we search out acceptable mechanisms to determine wise research directions.

Environmental Impacts

Although the potential benefits from the application of biotechnology to agricultural problems are enormous, significant potential safety problems need to be addressed. These safety questions are often coupled with other issues related to farming, such as soil erosion, the inappropriate use of pesticides, ground water pollution, and concerns for the sustainability of agriculture, among other issues. The public perceptions of risk from the applications of biotechnology clearly have impacted the process through which biotechnology research projects come "onstream." The public sector scientist is no longer receiving the special treatment once offered by society. Some federal and state agencies are now regulating the research process to ensure environmental and public health protection. To some individuals, this represents an ethical crisis because the values of agricultural scientists are publicly called into question, and new mechanisms for professional accountability are proposed.

The first U.S. biotechnology field test to be federally reviewed was that of Dr. Steve Lindow of the University of California at Berkeley. Lindow proposed to study a genetically engineered "ice minus" bacterium to protect plants from frost. The review of the proposal was resolved only after 4 years of federal litigation. The scientific validity of Lindow's test was never questioned. Instead, legal questions focused on the local and regional impacts of releasing the bacterium into the environment. The court decided that the Lindow study fell under the purview of the National Environmental Policy Act, requiring the proposed research to receive an environmental assessment before it could begin. This is considered as a major turning point for U.S. agricultural research and represents an enormous new burden placed on certain aspects of agricultural science.

Auburn University's Dr. Rex Dunham, a fish geneticist, has genetically engineered a rainbow trout growth hormone gene into common carp. He has asked for federal government permission to test the expression and inheritance of the gene. This is a fairly simple study, but in order to perform it, the test carp must be grown outside of a laboratory in experimental ponds. Dunham has developed a research plan that includes reasonable confinement methods. But for more than 2 years he has been trying to get federal approval to conduct the test. To obtain approval, Dunham submitted a detailed research plan that was reviewed by a USDA scientific advisory panel. Its evaluation and recommendation was used by USDA to prepare a detailed environmental assessment which concluded that the test would not significantly affect the environment. The USDA's document was then published in the Federal Register for a period of public comment. Meanwhile, Dunham's project awaits the system's response. Signs of frustration are evident on all sides.

These examples illustrate the growing distrust of science, particularly agricultural science, by some individuals and organized groups. National surveys indicate that public distrust is not widespread. The litigation is traceable to relatively few sources. As science and technology become more complex, public arguments are easily manipulated for ideological purposes. As Ruttan (1983a) points out, a response to questions of environmental safety through regulation places a burden on the scientists which may impede a desirable degree of freedom to conduct research on the unknown.

Heretofore, the scientific community has been treated as responsible and, in many ways, exempted from regulatory oversight. This has been especially true for agricultural researchers who have benefited from the public trust as non-vested scholars providing scientific services to the farm community. Has the recent shift in agricultural research partnerships, especially in biotechnology, caused a change in public perceptions? Has public sector agricultural research sold its credibility?

Whatever the reason, biotechnology has brought with it new demands for more accountability, more requirements for pre-experiment reviews, and costly permit/compliance regulations for what many in the scientific community now see as mostly imagined risks. The transactional costs (what it really takes to get permission) of the federal and state requirements for field testing genetically modified organisms has become an enormous burden. In a recent survey of U.S. biotechnology researchers, 12 percent of the respondents indicated that they had material ready to field test but chose not to bother because it would take too much time, too much paperwork, or because there is too much regulatory confusion (Ratner, 1990).

The situation thus becomes an ethical dilemma for the scientist who is committed to solving real-world problems, but is also frustrated by complicated and expensive regulatory burdens. Is this in the best interest of science?

Social Impacts

Biotechnology research offers a clear example of how a social issue can emerge through the implementation of new agricultural technologies. The U.S. is currently wrestling with the question of how far to go in preserving the American farm family. Part of the national effort to preserve family farms has involved the expenditure of millions of dollars to purchase cow herds to reduce milk production and sop up excess milk to maintain prices. Meanwhile, biotechnology has permitted economically feasible production of bovine somatotropin (bST), a hormone treatment for cows with the potential for increasing milk production by as much as 20 percent. The consequences of this and similar technologies are considered by some as doing more damage to the long-suffering American farm family.

At first glimpses, the bST example suggests that the solution to the social consequences is to stop research on technologies that would increase milk production. But this represents a new test—a "fourth hurdle" for product registration (the first three being safety, efficacy, and quality). This fourth hurdle goes far beyond previous tests to ask the question, "Do we need it?" The ethical implications of this question for agricultural research are enormous and not likely to be resolved in the near future.

Religious Implications

The scientific community has, in the past, been mostly excused from considering the religious implications of research, but those days appear to be gone. Biotechnological research creates new opportunities to intervene in natural processes in ways that seem, to some observers, to intrude on religious questions. Scientists can no longer continue to expect public support for research by simply asserting a philosophical commitment to learning the truth. New knowledge that offends the values of certain religious groups will engender political opposition that cannot be overcome by calls for academic freedom.

Responding to the Need for Ethical Awareness

As new relationships and successful partnerships among government, user groups, and the private and public sector scientific communities evolve, new processes to address ethical questions are needed. In the old "triangle," the feedback mechanism between agricultural researchers and farmers ensured a consistency with traditional agroethics. Today, public accountability of agricultural research is neither automatic nor ensured. Nevertheless, Ruttan (1987b) notes that "It is in society's interest to let the burden of responsibility rest lightly on the shoulders of individual researchers and research managers." This is not to say that regulatory procedures for research on human subjects, radiation hazards, biohazards, and other well-established programs of review and control over scientific research procedures should be abandoned. Although institutional and federal review and recommendations on these issues may be viewed as a significant nuisance to scientific research, such regulation seems necessary for today's complicated research projects. But if applied unwisely, regulation is not only a burden for scientists, it also can make scientific laboratories vulnerable to federal regulatory control in ways in which those who are unfamiliar with science do not realize. A delicate balance is needed in this area.

Implication for Researchers

Agricultural scientists must become advocates for the public interest by making a firm case for the benefits of scientific research. This requires a clearer vision of why particular research projects should be undertaken. Scientists must begin by studying the alternatives to a new technology, and become competent at making comparative judgments on the relative benefits and risks of undertaking research projects. In the future, scientists will be called on to provide an assessment of a technology before making choices. And they will be required to justify those decisions with arguments that clearly articulate the costs, benefits, and consequences of each alternative. How this will be done remains to be resolved.

Over the past few decades, disciplines in all scientific fields have become more and more specialized, including agricultural disciplines. As a result, fewer researchers have had the time or the necessary background to approach ethical issues with the same rigor that they approach scientific issues. The need for understanding ethics is so critical, and the problem has been unaddressed for so long, that some appropriate activity is needed soon. As a starting point, a series of workshops on ethics for agricultural scientists would help them better cope with the ethical issues now in the forefront.

Traditionally, most moral and ethical education has been delegated to families and churches. For whatever reasons, this "division of labor" no longer seems to provide a consistent set of principles to guide our behavior. This may be because our society is changing, and therefore, our ethics are changing too. We can no longer assume that we all share the same ethics, or that we all apply our shared ethical principles in the same way.

Because of the exodus from the farm, few agriculture students now have an agricultural background. Agricultural educators can no longer assume that students will embrace farm family ethics and moral values. If the demographic base for agriculture students (and future agricultural scientists) is no longer linked to a source of agroethics, will the American agroethical tradition continue to guide agricultural research? Probably not. Consequently, agricultural educators must teach agriculture students about these areas in which they have no experience, including ethics. Ethics education needs to be incorporated into curricula at the undergraduate and graduate levels, if not at the high school level.

Is higher education ready for this? Many of our agricultural faculty are not adequately prepared to teach moral values or ethics courses. Some may object to philosophers teaching ethics to agriculture students. Would they bog down with abstract ethical theories and leave students poorly prepared to address concrete problems in agricultural research and its applications? The educational emphasis must be on the practice and substance of ethics in agriculture, not the theory. This does not mean that ethics in agriculture can be taught without any ethical theory. What is required is a solid foundation in agricultural ethics based on appropriately selected agricultural case studies. Some minimum background in ethical theory would give students the basic skills required in analyzing case studies, identifying the nature of the ethical issues that emerge, and resolving the ethical consequences of alternative choices. At the very least, such an approach would make students sensitive to the ethical issues in agriculture and introduce them to problem-solving processes that could be utilized throughout their careers.

To suggest that we need to develop the ethical backgrounds of our agricultural students and scientists, however, is not to imply the content of that ethical background. The context should be determined by the relationship among traditional agroethics, the current state of American agriculture, and the interrelationships of these to the rest of American society.

For too long we have assumed that the values of agroethics will forever continue to underlie the farm experience. Accordingly, we have invoked these values to support our actions at individual and policy levels. Because ethics change over time, is it possible that the ethical base for agriculture has changed significantly, been lost, or been rendered obsolete? A major national ethics research program could discover the nature of the current farm value system, determine what remains of the traditional agroethical system, and help predict changes in values as they impact American ethical standards.

If agroethics survives as a guide to choices, it could serve as the foundation for teaching ethics in the agricultural sciences. However, agroethics evolved when the technologies being developed by contemporary agricultural research were inconceivable. Thus, can agroethics be expected to help us resolve all of the issues raised by new technologies? The challenge will be in adapting agroethics to resolve new issues and identify other, newer ethical principles to evaluate contemporary agricultural research. Regardless of whether agroethics survives, this nation cannot judge agricultural research based on our social values until we know what those values really are.

Ethics and Research Policy

Teaching and research in ethics may not be enough. Is this the time to start asking social value questions of research projects before they are initiated? These include questions like, "Who will benefit from the outcomes of this research and in what ways?" Fifteen years ago, grant submissions to the Pennsylvania Department of Agriculture required the principal investigator to rank order 10 "client groups" who would gain benefits from the proposed research. The client groups included the scientific community, farm community, consumers, and others. Today, this still seems to be a responsible question to ask of publicly funded research. But at first, this task was considered foolish and inappropriate for a research scientist. Have time and circumstances changed?

In the not-too-distant future, several questions will be asked of scientists before they undertake experimentation. The questions will require researchers to examine a research project's potential impact on the environment, the economy, society, and other areas beyond simply the development of new knowledge or better technology. This should serve as an honest exercise for the principal scientist. To determine this by regulatory law or delegate this task to peer review panels, as we do now for biotechnology research, are poor alternatives. Moreover, strong ethical values would provide more effective proper self-evaluation of research intent and benefits.

If scientists hope to "do it better," they need new methods to anticipate the expected impacts of discoveries and to cushion the negative consequences of emerging technologies. Withholding new technology from commercial application for a higher goal of social or environmental benefit is unreasonable. Although arguments can be made to the contrary, the track record of "technology management" clearly suggests that economic, not social forces, determine a technology's appearance or disappearance from the marketplace.

The current approach to protecting American farm families dispenses massive amounts of federal dollars as support prices to keep farmers in business, while market forces tend to drive them out of business. Ethically based arguments have been offered (Shrader-Frechette, 1988), but can we afford to create, as my colleague Dr. John Fulkerson calls them, "zoological gardens," to simply preserve a way of life? Can we not preserve our agroethical system as a scholarly pursuit rather than an economic war of wills?

This chapter has outlined an agenda that includes education, research, and a science-based evaluation of the impacts of technology on these ethical choices. This agenda should better address the needs of this historically important, culturally valuable, and socially accepted ethical system. Agroethics has contributed to America's greatness, wealth, and future in so many ways. It deserves our attention.

Notes

1. This "non-vested" confirmation of technology has evolved to become an important service provided by the public sector scientist for the farm community, especially for chemicals and seed. Most commonly, corporate research units sponsor field tests with small grants which cover a portion of the cost of the trials (typically $1,000 for a small test). The complications, however, are numerous. Public sector scientists who become broadly involved in such research often fail to undertake other, more publishable research. This can be ruinous for scientists seeking tenure or promotion. Some scientists accept invitations from corporate sponsors to go on fishing or hunting trips, or to attend lavish parties and other events as a guest of the company. This is a serious breech of the principles of disinterestedness. But too few institutions forbid such practices and the special standards for agriculture are lost among other values and practices of the rest of the university.

2

Balancing Moral Imperatives Through Rural Development

Vernon B. Cardwell

Introduction

There was a time when the role of the professional agriculturist was straightforward and singular of purpose. It was to feed the world; to get two plants to grow where one once did; to maximize production. We have made impressive gains in the yields of wheat, rice, and corn; pounds of milk per cow; pigs weaned per litter; eggs per hen; and feed conversion by pigs and chickens, but science and technology has not solved the problems of agriculture.

In fact, we have not resolved the problem of world hunger, and there is a growing awareness that our straightforward and singular approaches have created other problems. We have failed to recognize the interconnected complexity of the human activity system we call agriculture. Changes in one part of this complex system might have improvements for some, but may create considerably worse circumstances for others. What was an advantage today may turn out to be a disadvantage tomorrow. We have frequently examined only short-term benefits, not-long term, and we have seldom attempted to examine the ripple effect of a change in one enterprise upon the whole system because of the complexity and the existence of numerous "systems." Each farm itself is a system.

In recent years, the national perspective of agricultural research has gradually changed from mere production and economic functions to the more complex issues of agriculture's responsibility to the environment, to health, to rural communities, and to lifestyles. These new issues and concerns have challenged contemporary agriculturists. They have found that the very methods of determining the problems, of setting research priorities, and even the methods for researching have been found lacking.

Optimizing production can no longer be the sole concern of the researcher, technologist, or practicing professional agriculturalist. Societal pressure is focusing attention on questions of long-term sustainability of current production systems. Many are voicing increasing concern about damage to the physical environment wrought by current practices resulting in soil erosion and the biological consequences of practices upon genetic diversity. The intensification of many poultry and livestock enterprises has led to significant problems of manure disposal and environmental pollution. Furthermore, the degradation of water quality because of pesticides, salinization, and eutrophication, which is not confined to specific sites and spreads across neighboring ecosystems (legal and political boundaries) in a variety of interconnected channels, has become a concern of society as a whole.

Many farming practices reflect the external pressures to increase productivity (lower unit costs) and public policy (acreage allotments or bases, tax structures, etc.) for maximum short-term profitability. This has resulted in increased unemployment of rural labor and social dislocation.

In addition to constant cost/price squeeze, farmers are caught in a shift in other values. Concern about health-calories, cholesterol, nitrates, pesticides, etc., affect the demand for agricultural products and has created a new force in the marketplace: the consumer.

Urbanites concerned about conservation, natural resources, and animal welfare alter the traditional farm/farmer-oriented decision-making practices. Traditional uses of land for agriculture are being challenged by those desiring more wetland, urban zoning, and industrial parks with accompanying legislation restricting agricultural land uses.

The consequence of increased mechanization, a declining work force in agriculture, and concentration of the agricultural service sector has reduced the viability of many small rural communities. Thus, the decline of rural communities has become a concern of church groups, social action groups and local and state officials, resulting in rural community development programs in several states.

Historical Changes in U.S. Agriculture

The transition in American agriculture from the Jeffersonian self-sufficient yeomen farmer to the capital-intensive, technical business of today began well before the Civil War but was given significant impetus by the Civil War and the passage of the Morrill Act of 1862 creating land-grant universities in all states. The Morrill Land Grant Act was inspired not by the notion that agricultural practice and life needed dramatic correction or improvement, but rather the Act served to provide the children of farmers and laborers with opportunities similar to their more wealthy contemporaries, ie., for a liberal education, as well as agricultural and mechanical pursuits. The children would presumably return to create a better America, one in which they possessed the benefits of the more well-to-do (Marcus, 1987).

Federal funding for research in agriculture at the state land-grant universities originated with the passage of the Hatch Act in 1887. The Hatch Act provides "That in order to aid in acquiring and diffusing among the people of the United States useful and practical information on subjects connected with agriculture, and to promote scientific investigation and experiment respecting principles and applications of agricultural science..." (Kellogg & Knapp, 1966). The cooperative agricultural extension service with shared local-state-federal funding administered through the state land-grant university system was established by the Smith-Lever Act of 1914. These three landmark pieces of legislation created the teaching-research-extension organization, insitutional, and philosophy characteristic of today's colleges of agriculture in land-grant universities.

The wheels of change in agriculture began before the Civil War, but the war stimulated a pattern of mechanically induced changes in agriculture in America which continues with today's scientifically induced changes. Jordan et al. (1986) show how the rate of increase in agricultural productivity is correlated to sources of "power." From 1775 until the Civil War, U.S. agricultural productivity increased at a modest rate, correlated with "hand power." Productivity increased between 1865 and World War I with the introduction of "horse power." Between the world wars and extending to about 1860, "mechanical power" produced a sharp increase. Finally, from the end of World War II to 1975, Jordan et al. attribute the sharpest increase in productivity to "science power."

Mechanization

Mechanical innovations, beginning with the patenting of the cast iron plow by Jethro Wood in 1819, followed by John Deere's development of the steel plow (1837), made tillage of soil easier. The development of the first grain reaper (1833) and the first practical threshing machine (1837) set the stage for replacing manual labor with capital investment in the form of machinery. The first major economic incentive for mechanization came with the labor shortages of the Civil War. The second came with WWI and the widespread adoption of tractors. The

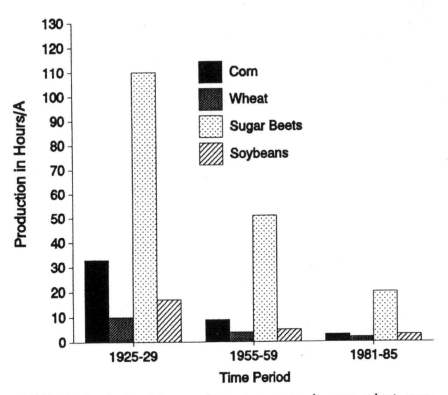

FIGURE 2.1 Production labor requirements per acre for corn, wheat, sugar beets, and soybeans in the U.S. for three select time periods.

expansion in agricultural productivity in the U.S. from the 1770s through the 1920s was primarily due to expansion in the amount of land under cultivation. Mechanization primarily influenced output per agricultural worker, not output per acre (Figure 2.1).

Science Power

Science power in agriculture began to have an impact in the 1930s with the adoption by farmers of hybrids corn even though the system for producing hybrids seed corn was developed by Donald F. Jones in 1918. This and subsequent genetic improvements were followed by the development of cheap methods of inorganic nitrogen synthesis, initially to support the war effort in the 1940s. At the end of World War II, the organic nitrogen capacity of the nation was diverted to agricultural use (Figure 2.2). The development of organic herbicides during the 1940s initially became available to farmers late in the decade. Through the 1950s, 1960s, and 1970s, the increased use of herbicides (Figure 2.3) and fertilizers correlate with the increased yield (Figure 2.4) during the period. Productivity increases in livestock followed similar patterns in crops, although at a different pace. Progress was greatest in poultry, followed consecutively by pigs, dairy, beef, and sheep. Productivity gains in animal production are possibly because of genetic improvement, equipment, feeding, and animal care products.

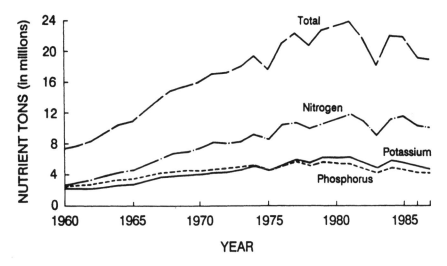

FIGURE 2.2 Total consumption of primary plant nutrients. *Source:* USDA, 1987. Fertilizer Use and Price Statistics, 1960-85. Statistical Bulletin No. 750. Economic Research Service. Washington, DC.

Themes of Concern

Agriculture is much broader than what occurs within the farm gate. Less than 2.0 percent of the population in the U.S. is engaged in production agriculture but more than 20 percent of the U.S. work force is engaged in the food and agriculture industry. Agriculture is a global industry responding to numerous market forces affecting demand, prices, food quality, and food safety. As an individual farmer, the margins for profitability for the production of traditional raw agricultural commodities are relatively small. Using average values based on national surveys, the per acre net returns above all variable and fixed costs, capital depreciation, taxes, and insurance for major crops for 1985-88 ranged from a negative $189 to $235 per acre. The poverty level of income for a family of four in 1988 was $12,091. To achieve net income above this level on small and mid-sized farms requires specialization in intensively managed, high cash value enterprises such as a dairy, where the return per cow ranged from $193 to $285. At $200 profit per cow, this

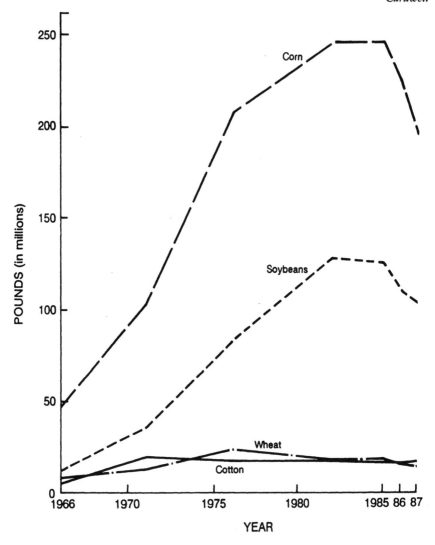

FIGURE 2.3 Herbicide use on major crops, 1966-1987. *Source:* National Research Council, 1989.

requires a herd of 60 cows. Not exactly a small farm or investment. "Beyond the large farm" implies there are unique problems for the not so large farms. However, the problems facing agricultural research are not restricted to production problems. The problems are not large farms vs. medium and small farms, nor specialized vs. diversified.

Farm Size

The output per acre from the 1930s until today is correlated with the combination of science power and mechanical power. This increased

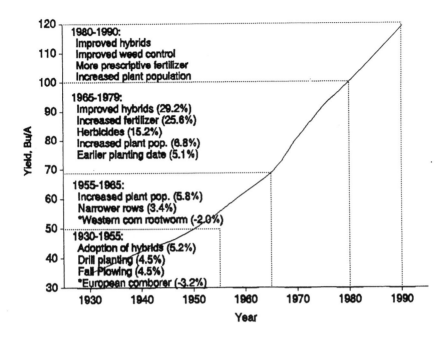

FIGURE 2.4 Technology and corn yields in Minnesota, 1930-1989.

productive capacity has led to surpluses and government programs in recent decades. The impacts of technology on agriculture since 1940 are evident, not only in productivity per animal unit, per acre, and per farm worker, but also in the number of farms and size of farms. Farm numbers and farm population in the U.S. peaked in 1933. The declining number of farms and concurrent decline in farm population results in an increase in the average size farm (Figures 2.5 & 2.6). These newer, larger farms are less self-sufficient, less diversified, and more reliant on machines and chemicals than were their predecessors. They are more capital-intensive, market-oriented, and are users of new technology. With the rapid changes in the structure of agriculture in the 1980s, many questions are being asked about the future direction of agricultural research.

Are we reaping the harvest of the seeds of change sown in 1862 with the establishment of the land-grant universities? What changes are needed in our agricultural research and extension/teaching programs to alter the course of agriculture for the future to meet current and future societal and agricultural concerns? If there is one constant in the last 100 years of U.S. agricultural history, it has been change! From 1930 until 1970, the ratio of farm labor wage costs in relationship to

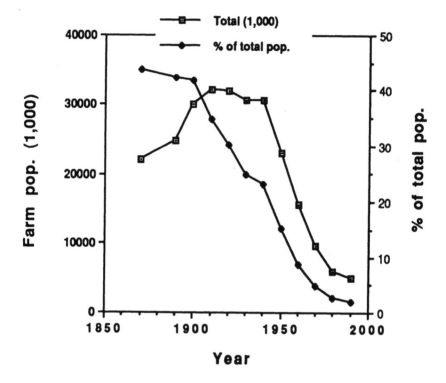

FIGURE 2.5 Farm population and percent of total, 1880-1990. *Source:* U.S. Bureau of the Census data.

machine costs increased 2.8 percent per year. The higher real wages available in the non-farm sector resulted in substantial substitution of mechanical power. The removal of the labor force also freed land resources through the prompted substitutions of larger tractors and machines for draft animals and labor. Biological inputs enhanced production per acre, but not farm size per se.

The Office of Technology Assessment (OTA) (1985) notes that some of the principle advantages of very large farms are not technological, but are buying and selling in large quantities and access to credit. Among various income tax provisions, interest payment write-offs had the greatest effect on farm growth rates. Capital-intensive farms benefited disproportionately from depreciation allowances and investment credits. Thus, the high tax rates of the 1970s created incentives for farms to expand for tax reasons. Tax credits stimulated farm growth by increasing the funds available for investment. Analysis of the benefits of the 1981 tax act suggests that tax savings become more dramatic as farm operator incomes increase.

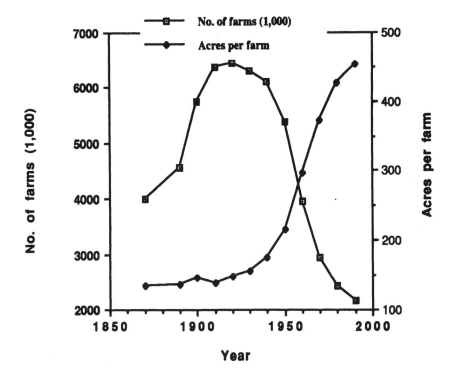

FIGURE 2.6 Number of farms and acres per farm, 1880-1990. *Source:* U.S. Bureau of the Census data.

U.S. agricultural research clearly facilitated, though it did not dictate, increases in farm size and fewer farmers. The work of Hayami and Ruttan shows that technology is less important than financial forces in explaining structural changes in agriculture. In the past, farm size increased because the opportunity cost of machinery was less than that for labor, but in the mid- and late-1970s this changed, and the rate of farm size increase decelerated. Financial burden, rather than technological problems, beset the farmers in the late 1970s and early 1980s.

Sustainable Agriculture

Buttel (1987) argues that one of the major types of criticisms of the land-grant system during the past 5 years has been criticisms from farmers. A survey conducted by a land stewardship project of farmers in Western Minnesota found that farmers were most concerned about erosion effects on water quality, pesticides, and government programs. The lack of research on "low input" systems and research on bovine growth hormone have caused a stir among Minnesota and Wisconsin farmers (Kelling, 1989).

As with all professions, respect, honor, pride, and dignity must be associated with the profession. When the circumstances of farmers place them below poverty income levels and the desired options open to farm family members are less than those in other segments of society, then the respect, honor, pride, and dignity associated with the profession becomes suspect. The farm and rural area problems of the early- and mid-1980s represent a situation in which non-production problems were driving the system. For example, growth in real interest rates, a strong U.S. dollar, restricted money supply, and low inflation all contributed to make U.S. agricultural products expensive, thereby resulting in reduced exports. This was coupled with a precipitous drop in land values, farm prices and farm incomes. Profitability in agriculture was significantly reduced, and the perceptions of farmers and non-farmers were that agriculture, as practiced during the last 20 years, was not desirable. The result has been a groundswell of interest in "sustainable" agriculture. Sustainable agriculture has become a banner under which environmentalists, individuals concerned about lifestyle, methods of crop/livestock production, and concern for rural life and communities have come together.

Environment

Environmental issues now have a permanent place in our political life. The economic benefits are far more quantified and measured than are the social and health costs of growth associated pollution. Thus growth has an unnatural edge over environmental quality in decisions on technology-related growth.

Environmental groups have shifted their focus from the right to clear air and water, to a focus on human health and well-being. Borelli (1988) indicates the effect has been to quadruple to approximately 25 million the number of people involved in local and regional environmental causes. Eco Net, a highly sophisticated telecommunication system serving the global environmental community, has more than 190 organizations using the system.

Castle (1970) notes that the agricultural research community has been preoccupied with "just increasing agricultural production." We have been singularly effective in our direct efforts, but we have largely ignored the indirect or secondary effects of our efforts. The social problems we now are addressing, in large part, are a manifestation of these indirect efforts coming to the surface.

Risks

Peter Sandman of Rutgers University makes a distinction between hazard and outrage in the conflict between expert versus public perceptions of risk. To the public, risk means a lot more than mortality statistics which might be classified as "what are the hazards?" Sandman (1989) suggests that everything else that goes into the lay public's risk perception is "outrage," i.e., everything about risk that is relevant except how likely it is to hurt you. Perceptions, people's concerns, beliefs, and fears are more a product of outrage than hazard. Understanding these perceptions will help us establish some of the criteria for setting research priorities.

Sandman discusses 12 major variables of perhaps 25 identified variables relating to outrage.

1. **Voluntary versus involuntary:** Involuntary or coerced risk is perceived as more risky, even though it is no more hazardous. Voluntary risk correlates with acceptability.
2. **Natural versus artificial:** A natural risk, such as radon, is more outrageous than a voluntary risk but less than a forced risk. In this case, there is no enemy to get mad at.
3. **Familiar versus exotic:** Familiarity leads to underestimation of risk. The more exotic, the more unfamiliar, the greater the potential for sense of insecurity or outrage.
4. **Not memorable versus memorable:** Association with memorable disasters or accounts, e.g., Love canal, Bhopal, *Silent Spring*, increases the perception of risk. Symbolism becomes very important in managing perceptions.
5. **No dread versus dread:** Perception of environmentally induced asthma and emphysema problems are less dreaded than cancer. More concern about air and water than about food.
6. **Knowledgeable versus unknowledgeable:** The public prefers a known risk which can be measured rather precisely to an unknown risk of smaller magnitude but which cannot be measured precisely and has a large margin of error. The latent and often untraceable causes of cancer are a major cause of alarm.
7. **Controlled by the individual versus controlled by the "system":** Persons who are in control, whether driving or running a meeting, feel there is less risk than when someone else is in charge, e.g., use of household pesticides versus community-wide public spraying for pests.
8. **Fair versus unfair:** Coerced risk is considered unfair. The more sharing of benefits (financial or otherwise), the more acceptable the risk or the lower the outrage.

9. **Morally irrelevant versus morally relevant:** Some risks are real hazards; others are wrongs—financially, socially, and ethically. The moral imperatives will be obeyed without counting the cost. Cost-benefit analysis is not persuasive.
10. **Affects the strong versus the vulnerable:** This is related to moral relevance. As a society we believe that the vulnerable deserve special protection.
11. **Trustworthy sources versus untrustworthy sources:** The relationship with the person, company, or agency imposing the risk, i.e., patient-doctor relationship, student-teacher relationship, or farmer-extension agent relationship. There is a growing sense of public distrust for chemical companies and for many of the state and federal agencies working with these groups. Trust is built, not demanded.
12. **Good process versus poor process:** The process by which industry and agencies interact with the public influences the outrage.

- Openness versus secrecy: openness wins.
- Acknowledging versus denying or ignoring problems.
- Acknowledging that people can understand complexity.
- Courtesy and caring versus annoyance and defensiveness: Big brother knows best but doesn't cut the mustard.
- Shared values and preference: the more similar our personal attributes the less suspect we are.

Avoiding outrage is the only way to maintain public confidence and support. As Sandman points out, the "job of risk communication, in other words, is to produce a public that has the knowledge, the power, and the will to assess its own risk rationally and which ones it ought to tolerate and which ones it ought to eliminate, and act accordingly."

Lobbying Groups

The membership rolls of national environmental interest groups have grown in recent years due to the influx of mainstream members of the American public who are increasingly concerned about the activities of modern society upon the quality of our environment. Many are national organizations and do not reflect other national, as well as state and local groups, working for a better environment.

There is general consensus that policy is set by bringing a consensus to centralized points of decision-making. Within the agricultural establishment, highly centralized clusters or networks of participants often develop their own agreements in a fragmented and decentralized universe. The proliferation of numerous special interest groups tend to disrupt the networks, minimizing the opportunity for consensus and reciprocal agreements, which adds to the confusion for policy-makers as they seek to agree on program content and funding. Browne (Hadwiger & Browne, 1987) identified 215 interest groups with positions on federal agricultural research. Five types of concerns were identified: (1) questions of cost, (2) program beneficiaries, (3) product and safety, (4) environmental use, and (5) control of research agenda.

The potential consequence of these competing forces in a political climate where few additional new federal research funds are likely is the redistribution of available funds among a larger number of targeted clientele. In this way, the critics could gain their own selective form of benefits. Such a redistribution would have a restrictive impact on present research efforts. This, in turn, could lead to the further diminishing of the credibility and perceived impact of agricultural research organizations, and allow policy-makers an excuse to make cuts in what they see as controversial programs.

Congressman George E. Brown, Jr. (1989), chair of the subcommittee on Department Operations, Research, and Foreign Agriculture of the House Committee on Agriculture points out that:

> in the midst of confusion, public interest groups have been calling for increased attention to food safety, farm worker protection and the environmental impact of agricultural production. They are not interested in hearing about higher yields or new pesticide efforts when the news is about high surpluses or pesticides in drinking water and food. Public perceptions shape the political reality, whether they are scientifically accurate or not.

Ruttan (1987) suggests that agriculturalists have been and are reluctant revolutionaries. "They have wanted to revolutionize technology but have preferred to neglect the revolutionary impact of technology on society." Research managers have a clear responsibility to inform society of the impact of economic policy on:

1. The choice of mechanical, chemical, and biological technologies by farmers.
2. The incidence of technical change on the distribution of income among labors, landowners, and consumers.
3. The structure of farming and rural communities.
4. The health and safety of producers and consumers, and the responsibility to enter into intellectual and political dialogue to gain

convergence between market prices and total social costs, which includes all hidden or deferred costs such as environmental degradation, and to integrate individual and societal preferences.

Agriculture as a business was initially reflected by the plantations of the South which focused on export crops, and later in the 1870s-1890s on the bonanza farmers of the Upper Midwest and the Far West. Agriculture as a way of life was replaced with the desire to rationalize operations and to conduct affairs as efficiently as possible to reap the greatest profit. The emphasis on agricultural production led to a corresponding emphasis on agricultural research. The research system had originated from a notion of the desirability of increased agricultural efficiency. Productivity would help farmers. Rationalizing farm operations would put agriculture in line with other American industries and provide farmers and their families with amenities similar to their city dwelling contemporaries.

The Hatch Act of 1887 institutionalized the agricultural experiment stations by providing funds ($15,000) for each state and territory for experimentation and investigation, and recognized the land-grant universities as the site for this research.

Two principal actors have been involved with the public agricultural research supported since the passage of the Hatch Act in 1887: farms and agricultural scientists. Prior to the 1870s, farmers did their own research, which focused on developing and improving machines to make their way of life easier. Scientists, as a part of the academic community, sought to develop new knowledge and understanding. Legitimacy of agricultural research leads to support "in the public interest." Client interests, producers, and consumers, who are direct beneficiaries of research also provide a base for the material support of agricultural research. In the past, consumer groups who were the largest beneficiaries have not significantly impacted resources for agricultural research primarily because (1) consumer benefits are diffuse rather than exclusive; (2) research institutions cater to producers; and (3) consumers are not well-mobilized.

Producers benefit as first users of new knowledge, and as the new knowledge is broadly adapted and adopted by other producers, consumers become the primary beneficiaries through improved quality and reduced prices. Producers do not support research in general, but rather a particular set of research activities (i.e., specific commodity-focused initiatives).

Farmers, through commodity groups and other organizations, agreed with a common goal for agricultural research: to improve efficiency.

This has been a unifying goal for farmers, researchers, and political groups until the last decade.

However, as the political clout of agriculture has declined, voices of discontent have been heard. Hadwiger and Browne (1987) identify the following categories of complaints, and those who make them:

1. Research benefits are given to those producers seeking research and to the researchers themselves, rather than to farmers generally. This complaint arises from public interest groups who want benefits for small farmers, consumers, and others.
2. Products developed by researchers are not always safe and nutritious, say consumer groups.
3. Food industry groups complain that the products are not always developed to be most attractive to consumers.
4. Modern agriculture adversely affects the natural environment, say environmental groups. Economic development agencies add that agricultural pollution also impedes the development of recreation and tourism in states in which those industries are competitive with agriculture.
5. Private agricultural researchers complain that public funders are released into the public domain, providing unfair competition with those who seek to profit from knowledge.

Extension

The Smith-Lever Act of 1914 created a network of federally supported county extension agents, under the umbrella of the land-grant universities, who work to bring the fruits of agricultural research to the farmers.

Crowder (1987) reports on interviews with 41 extension agents and 90 farmers in Imbabura Province, Ecuador. He found that 48 percent of the agents thought the larger farmers were most likely to adopt new technology and practices, followed by 38 percent of medium-sized farmers, and only 14 percent of small farmers were most likely to be innovators. The agents, in turn, spent 36 percent of their time with the large farmers, and 41 percent and 23 percent respectively, with the medium-sized and small farm operators.

The Wisconsin Agricultural Experiment Station and Cooperative Extension Service conducted a series of five sustainable agriculture "listening" meetings in early 1989 to "seek ideas and suggestions from interested citizens for future research and extension programs in sustainable agriculture and related areas." Kelling (1989) notes that "diverse interests have rallied around the banner of sustainable agriculture, including environmental resource protection; product quality and con-

sumer safety; profitability of individual farms; policy questions related
to competitiveness of medium-sized family farmers; and the viability of
rural communities." The meetings involved 231 participants who shared
ideas for 20 areas of research and 13 suggestions for improvements in
extension education. Each suggestion was identified at least three
times for a listing in the final report. The extension criticisms surfacing
most frequently were:

1. Many participants perceive extension as a top-down, education
   organization.
2. As non-users of extension and university, there was a lack of
   knowledge by this clientele group about what extension and the
   university had to offer.
3. A substantial number of individuals in the audiences have chosen
   to farm organically or with reduced inputs. Can extension serve
   the needs of this group while continuing to serve the needs of
   other groups?
4. Extension should become "advocates" for sustainable systems.
   Structural trends in U.S. agriculture continue to conflict with developing alternative approaches. Youngberg (1987) suggests that
   for the most part, structural and technological trends have impeded the development of diversified, biologically based, resource-
   conserving farming systems. Energy and capital intensity, level of
   specialization, and a trend to larger size are characteristics of an
   industrial form of agriculture. This form of agriculture is supported
   by a public agricultural tax, credit and commodity policy, and
   competitive character of farmers caught in a cost-price squeeze
   and operating with short-term planning horizons.

Current Public Perception of Agricultural Research

Critics of the land-grant university system agree that the primary
clientele of the land-grant research/extension system are large farmers,
agribusiness, chemical and machinery corporations, and other non-farm
interests. These clientele leverage the use of university faculty time,
laboratory or field resources, and general overhead, which benefits the
larger farms or businesses, not the moderate-scale family farm.

Critics also allege that researchers ignore the ecological impact of
current practices and no incentive exists within the system to find
alternative economically and ecologically sound methods that have
minimal social impact.

There are issues in which agriculture has become embroiled which
cannot be adequately addressed by traditional production agricultural

research methodology. An example of this is the often vociferous dialogue on sustainable agriculture. Many hours by hundreds of people have been spent attempting to define "sustainable agriculture." The point is that we have many situations which are ambiguous and laden with conflicting facts, views, or perceptions and philosophies.

Clarifying values becomes as important as the underlying goals in setting research priorities. Agricultural scientists are generally not prepared to seek improvement in complex, messy, and dynamic problem situations in which there are multiple right answers and no obvious optimum, and the best outcome is an improvement in the situation.

Research organizations are engaged in conflict with the public over a series of questions. What is appropriate research? Is the research scale neutral? Does the research benefit one group of farmers at the expense of other groups? These difficult questions can only be resolved as we take an interpretative intervention approach. This approach appeals to the perception of fairness, and responds to various moral viewpoints, outlooks, and frameworks by hearing out each viewpoint, by teaching us how to live with our differences, and by searching for common group understanding. We must learn not to presume.

If current research is perceived as oriented toward the large farm, then we must establish mechanisms which provide a vehicle for research goal setting:

1. Where all participants perceive themselves to have been treated fairly and impartially.
2. Where the treatment is symmetrical and even-handed, and where turnabout is fair play.
3. Where participants believe they have been understood and their views respected and impact incorporated into the final outcome.
4. Where participants feel they were a part of the designed resolution and that the final outcome was not imposed from external interests.

Shepard (1984) notes that "scientists share a collective responsibility to accurately inform the public of the results of scientific investigations that are relevant to public controversy. Since transcientific (factual data and value assumptions) questions are involved, however, the scientist who rigidly segregates facts and values will be unable to meet this responsibility."

If we view agriculture as a "human activity system" and fail to recognize that morality viewpoints other than our own exist, we will be unable to respond with moral sensitivity and in good conscience, respecting other views and the complexity of various situations. But by lowering our sights from conquest (win-lose) to coping (win-win) we can adapt and mobilize our moral sensibilities to better influence the present course of events. As agriculturalists, we must be aware that problems regarding values are here to stay, and it is important that agricultural scientists set an example of coping well with a healthy plurality of moral concerns.

Shepard (1989) notes that critics of the agricultural research establishment have been preoccupied with the "values" of conventional agriculture, with diagnosing the "ills" of conventional practices, and with prescribing alternative values and practices. The "establishment" defended agricultural science by appealing to the authority of the scientific method as value-neutral and above politics and morality. Resolving the conflict requires presenting the context of the controversy in a manner that relevance to the issues at stake can be grounded in interpretations of the issues that are acceptable to all the contestants. What is the common ground between scientist and non-scientist? Between factual and normative considerations? Between opposing viewpoints? Between divergent meanings and conflicting standards of relevance? What is credible in the way of facts depends on what is credible in the way of values and vice versa.

Bonnen (1983) finds "changes in society's values and social agenda, in part the consequence of externalites to agricultural policy and production, will remain an important source of disequilibria. This will require not only social, physical, and biological science, but also humanities research on the ethical and value conflicts in the choices that must be made." Vogeler (1981) estimates that no more than 35 percent of U.S. farmers have benefited from research performed in the publicly funded research sector.

Science has been considered valuable because it seems to lead down the objective path to truth or knowledge, with scientific "facts" considered as units of knowledge. Conventional science has tried to separate intuitive (experience) knowledge and context from scientifically described knowledge. Scientists tend to measure variables that are quantifiable and avoid problems which cannot be easily quantified, referring to them as "nonresearchable." This prevents exploring the relationship of different phenomena and the interactions created when socio-economic, cultural, political, and emotional context of real-world situations interact with scientific fact.

Agriculture as a human activity system is value laden. Values held for the land, lifestyles, rural communities, the economic system, and the environment all influence the decision-making process.

Lockeretz (1988), in discussing the issues of sustainable agriculture, raises the question, "Does understanding sustainable agriculture involve concepts that are fundamentally different from conventional (agricultural) systems or do we need only to extend the application of known princi-ples to the conditions that prevail under sustainable practices?" One could insert mid- and small-sized farms for the word "sustainable" with a similar set of responses. For example, both large and other farms can be described or analyzed using the same concepts and differ only in the specific condition created by the differing situation associated with size. An alternate view might be that the complex interaction of integrated components of any farming system create new phenomena which emerge as one examines each unique system and cannot extrap-olate from one system to another.

Lockeretz (1988) also poses the question for agricultural research policy-makers: Are established methods of conducting agricultural research appropriate for all farms? Perhaps all that is needed is some "fine tuning" to ensure applicability to broader situations. Johnson (1984) and Madden (1987) espoused the idea that a "new covenant" for agricultural academics is needed, e.g., set aside our personal chauvinism, and seek more effective combinations of interdisciplinary problem-solving and subject matter research to obtain societal goals.

New Research Approaches

Other people believe new approaches and methods are needed, in-volving a new paradigm with different theories and concepts. Three have been suggested: (1) agroecology, which draws more heavily on the principles of ecology (Conway, 1985; Altieri, 1988), (2) the FSR/E-farming systems research and extension methods. (Hildebrand, 1986; Norman & Collison, 1985) and (3) systems agriculture (Bawden et al., 1985) (Wilson & Morren, 1990).

Conway (1985) describes the agroecosystems research approach as not only examining productivity but also explicitly examining other indicators of performance—stability and equitability—and the tradeoffs between them.

FSR/E has focused on limited resource farmers. It has sought to develop a basic system for research and extension with the strong involvement of the farm families themselves in all stages of research and development with the objective of increasing the development of appropriate technology suited to a group of homogeneous farms.

Systems agriculture recognizes that agriculture is a human activity system with multiple interacting forces, including social, economic, po-litical, scientific, and technical issues. Solutions to these messy situations are not found in optimizing functions that increase productivity, but in examining the whole system in which solutions are found that are satisfying to the owners of the system. The focus of traditional and some FSR research leads to the purpose of improved productivity. The agroecosystem and systems agriculture approaches do not assume that the purpose of the research is to increase productivity. Increased productivity is not synonymous with increased production.

Suppe (1987) suggests a number of reasons why traditional agricultural research is inapplicable in on-farming situations, including:

1. Pure or basic-oriented rather than applied.
2. Research outcomes are not cost-effective.
3. Research utilized seeds, feeds, and other materials not locally available.
4. Local field conditions differ from research sites.
5. Poor research.
6. Research and recommendations do not correspond to the farms' goals.

Suppe, a professor of philosophy and part-time farmer, presents the process for decision-making on his farm. First, he utilizes that quantitative data of agricultural research found in professional journals. He suggests that farmers not associated with a land-grant institution still have access to the technical literature, albeit somewhat select, in the form of farm magazines, and through extension short courses and research field days. Second, he uses informed expert opinions, such as various public and private professionals, and other well-respected farmers.

Dick Thompson of Boone, Iowa, has become a nationally recognized farmer whose opinions are highly valued by other farmers as illustrated by attendance at field days on his farm. Every community has such respected farmers. Third, Suppe synthesizes all of the information as it relates to his farming situation. The process does "not yield probablistic expectations," but rather it "is a qualitative plausibility assessment." He argues that this approach takes into account multiple factors, and these are more fine-tuned and reliable than the use of probablistic results of a single study. This decision-making process does not depend on a set of rules and, as such, becomes somewhat of an art which is enhanced by the insight and experience of the person making the decisions.

One of the difficulties of trying to improve situations for mid-sized and small farmers is that each farm represents a system, and with any kind of negotiation, farmers cannot negotiate with abstractions because the decision process is at the personal level.

Busch and Lacy (1986) cite forces that are creating a need for changes in approaches to agriculture and new technologies.

1. Changing food consumption patterns.
2. Increasing concerns about the continued availability of soil and water resource bases.
3. Changing demands of domestic and foreign customers for forest and range products.
4. Increasing public concern about the effects of scientific, economic, and technological developments on human health and quality of life. How do we evaluate low-level carcinogenic risks from multiple sources? How do we evaluate the safety of genetically engineered microbes that may be used in food processing or production?

Schweikhardt and Bonnen (1986) point out that agricultural research policy was once concerned primarily with allocation resources, but now it involves a growing list of new issues and conflicting interest groups. Since its inception, a key problem for experiment station research has been priority-setting, and E. Lewis Sturtevant, of Cornell University Experiment Station, notes the Station determines "what the farmer wants to know" and "why he wants to know it."

All decisions involve both normative (values) and positive (factual) knowledges. The public issues related to agriculture being debated today tend to be poorly understood by scientists whose emphasis on facts makes encompassing values into decision-making very difficult.

Today, the private sector accounts for about two-thirds of all agricultural R & D expenditures (Ruttan, 1982). Two-thirds of the private sector and 15 percent of the public sector R & D expenditures are concentrated in physical science and engineering. Three-fourths of the public land-grant and federal research, and 20 percent of the private sector effort is devoted to biological science and technology. The social sciences received less than 10 percent of public R & D funding, and less than 3 percent of the private funding.

Bonnen (1987) notes that academia's response to our expanding knowledge base has been to form more specialized disciplines. This has created increasing organizational distance between applied researchers and the basic science disciplines. Furthermore, academic science frequently treats activity that is not biological or physical and quantitative as subjective, and thus nonscientific. Therefore, questions about values are not considered.

In the past, the common criteria and, in some cases, the sole criteria for selecting research projects was whether they promoted efficiency. Madden (1987) points out that side effects were society's responsibility to assess on the basis of costs/benefits. Madden suggests that researchers and research institutions should commit themselves to "scientifical-

ly excellent, socially relevant, and ecological responsible" research as
we seek to avoid undesirable and unwanted side effects.

The search for academic excellence has divided agricultural professionals who strive for peer recognition within academia versus problem-solving and service to people. Agricultural universities need to
bring an integrated focus on both the agenda of science and the problem
solving in agriculture.

Systems Approach: An Example

The Redwood County Minnesota project, partially funded by the
Otto Bremer Foundation, was designed as a systematic approach to
examine problems of agricultural and rural communities. The project,
initiated in 1985, attempted to determine the issues and concerns of
farm families through open-ended oral interviews conducted in the
family home by teams composed of an agricultural scientist (agronomy,
horticulture, or soils) and social scientist (economics, anthropology,
sociology, or family social science). The interviews, lasting 1 1/2 to 2
hours, attempted to identify: (1) history of ownership and operation of
the farm; (2) current farm situation, i.e., size, acres owned/rented, crop
and livestock enterprises, nature of farm records, labor, etc.; (3) the
most pressing problems of the farm, the family, and the community; (4)
future changes or goals for the farm, the family, or the community; (5)
information sources and basis for making decisions; and (6) assessment
of problems in the community affecting agriculture and family life.

The initial survey involved 13 faculty and graduate students inter-
viewing 61 farm families (approximately 3 percent of the farm population of the county). Given the date of the survey (July-August, 1986),
the farm economy was just beginning to recover from the financial
crisis of the previous 3 or 4 years. The foremost concern was agricultural
prices and profitability. Other major concerns among the families
revolved around church, schools, and community, followed by family
concerns about health, non-farm employment, and intergenerational
transfer of property. Government farm policies and programs, commodity marketing, pollution, pesticide health risk, banking regulations,
land rental agreements, and farm records for non-tax purposes were
the next most important concerns. Commodity-related concerns were
at the bottom of the farm family list of concerns.

The systems approach to problem identification demonstrates that
the top priorities of farm families are frequently complex agri- and
socio-economic problems, which one would expect if we consider agriculture as a complex human activity system and not just a commodity
production system.

Many of the problems identified call for group efforts (church, schools, communities, government programs, and banking regulations) to improve the situation. Other concerns have individual impacts in which researchable problems could yield alternative solutions appropriate to the individual farm families (systems), e.g., land rental agreements, farm record systems, or commodity marketing programs. Respondents did not issue a significant plea for commodity research per se.

The oral survey provided several benefits to the researchers. It was humbling to find out that their passions were not the hot issues of the farm families. It was rewarding to see people share a general concern for their community. It was stimulating to work with colleagues in other disciplines who have a different mind set.

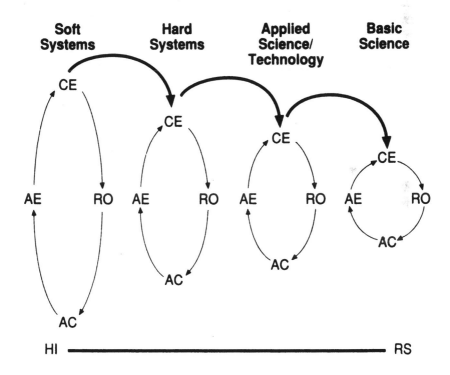

FIGURE 2.7 The holistic integration-reductionist separation axis: a spiral of interconnected learning cycles. *Source:* Wilson & Morren (1990). Reprinted by permission.

The complexity of the problems and issues identified make it clear that problem resolution using traditional scientific methods of fact finding and optimum level seeking are insufficient. The need to operate at all levels of a learning (finding out about) spiral (Figure 2.7) was evident. Many of the issues identified by the farm families were soft system issues. For example, the problems and goals are messy and complex and not easy to define because there is no general agreement on problems, goals, or purposes. The research method must facilitate an inquiry process, whereby the nature of the complex problematic situation is articulated and an improved or more satisfying future state is agreed upon by the owners of the problem. This methodology is common among social scientists but seldom in the tool bag of skills of agricultural scientists.

Many of the issues identified by the farm families require this approach to improve issues about church, schools, communities, government programs, environmental quality, and land use. These issues are qualitative and require the researcher to be keenly aware of the boundaries of the system, to understand the relationships and interactions of components, to be concerned with "what could be" rather than "what is" and to combine the feelings, beliefs, and values of the client with the pragmatic physical and biological constraints of the situation. Systems thinking implies a more holistic or broader perspective, encompassing humanistic and scientific entities.

Hard systems are best suited to tackling problems involving biological, physical, mechanical, and human activity systems for which clearcut quantitative goals can be established. As with soft systems, hard systems have defined boundaries, inputs, outputs, essential transformation, and some measures of performance. Mathematical relationships are typically sought and measures of efficiency are the usual criteria for evaluation.

Applied research in agriculture is often a specific farm, field, or animal problem in which the question raised is, "How can I improve the effectiveness?" The problem is usually a subsystem, a component of a larger set of enterprises or activities. Seldom is the problem viewed as a "people" problem or a system problem.

Basic research revolves around finding out about reproducible "facts." It is puzzle resolution. It answers questions of "given this phenomenon, why is it so?" The scientific inquiry and protocol which is the basis of our scientific agriculture associated with basic research has generated the fundamental principles to explain the natural phenomenon.

The fact that all agricultural phenomena have not been explained and every little aspect is not predictable attests to the complexity and interdependence of agricultural enterprises and activities and argues for all methods of inquiry—the scientific, applied, hard systems, and soft systems—to address the problematic situations in agriculture. A systems perspective appears essential to finding new ways of learning about and setting strategies for agriculture. The Redwood County project has attempted to embody all levels of learning about the problems and issues in Redwood County farm families.

At the soft systems level, a part-time county rural community development person was hired to address church, school, and community issues. People have organized action committees in the three communities with which we have worked. There is no optimum, or best solution. Each community has gone in different directions, but each has a greater sense of satisfaction and belief that their circumstance is better as a result of the mediated efforts.

At the hard systems level, the issues that seemed researchable or doable, given the resource and time constraints, were educational programs to provide individual farmers with the capacity to function more effectively in the larger agricultural system. In the 1986 survey, farmers had identified commodity marketing as a concern. It was ascertained that the core of the problem was the lack of a working knowledge of all of the instruments and techniques available to farmers in marketing their commodities. Farmers had heard of "futures," "options," "hedging," "calls," "puts," etc., but most were non-users of the market instruments because of the "horror stories" of people who had "played the market." Playing the market and using the instruments in the marketplace as a part of a farm management marketing strategy are two separate and distinct activities. To overcome the problems of understanding, marketing clubs were started in each of the three target communities. The clubs were initially established with strong support from the county agricultural extension office and several local commodity brokers and elevator managers. The clubs have become self-governing and two are self-sustaining after two years of operation.

On-farm applied research has explored the issue of how best to utilize farm produced manure for a group of livestock producers. The common practice in the county is to spread the manure and then apply commercial fertilizer in a blanket application, giving essentially no fertilizer credit to the applied manure. Most farmers have not had their manure analyzed for mineral composition, nor did they have a very accurate estimate of the amount applied per acre. Five farms were selected to determine the amount of manure applied, to analyze the manure, and to compare corn response to no fertilizer or manure, manure only, commercial fertilizer only, and the combination of manure and commercial fertilizer. Variations in manure quality were identified due to methods of handling and sources. Variations in response to applied commercial fertilizer were shown, due primarily to the amount of manure applied and previous history of manure application. Even at moderately high yields 120-155 bu/A, some farmers appeared to get little or no response to applied nitrogen, suggesting the potential of over-application and associated ground water problems. This activity was an effort to help farmers fine tune their program and make more effective use of their available resources.

Basic research efforts were directed at investigating a new and potential alternative crop for the area. Amaranth has been suggested as a new crop. There is little or no information about cultivars; date, rates, and depths of planting; row spacing; and weed control methods. Because of the exploratory nature of the efforts, the number of variables involved, and the need to meet thesis research standards, plots are located on research farms in the county and at other locations in the state.

The Redwood County project has given us the opportunity to look at agricultural problems at multiple levels, from macro to micro, from sub-systems to macro-systems, and from biological/physical field situation to the family and community-based situations. The project has helped develop the perspective that agriculture is a dynamic and complex human activity system, requiring multiple strategies to improve the situations.

Conclusions

The history of U.S. agriculture is a history of change. Initially, the exploitation of land was followed by the impact of science and technology, in part enabled by the Morrill Act of 1862, (land-grant universities), the Hatch Act of 1887 (agricultural experiment stations), and the Smith-Lever Act of 1914 (agricultural extension services). These enabling pieces of legislation provided the intellectual capital to develop and exploit the biological and physical resources creating one of the world's most productive and efficient agricultural systems. This, and other legislation, provided the institutional and political structures to support the research and production systems, and to stimulate the entrepreneurial efforts of agribusiness.

The power base for generating support to continue the structure and resource development of agriculture has declined with the loss of farmers and rural populations. The agricultural lobbying group and their coalitions have shifted, resulting in an evolving redefinition of the mission of agriculture and agricultural research. Productivity and efficiency will continue to be a goal of agriculture, but new practices must be ecologically responsible and socially relevant.

Agriculture must be viewed as a human activity system in which the methods provided by research are given meaning by the various clientele groups—where successful research reflects a more satisfying situation. Research efforts of the future must be a balancing act of ensuring adequate productivity, profits, and efficiency, while sustaining a favorable environmental and societal impact.

3

The Value Measure in Public Agricultural Research

Jeffrey Burkhardt

Introduction

This chapter describes and analyzes the primary philosophical and ethical issues and arguments associated with the impacts of mainstream U.S. agricultural research and the technology development agenda (hereinafter, referred to as the "research agenda") on family farms. The focus is on those arguments or positions that seek to establish or reaffirm a legitimate role for what can be called "family farm research" in the research agenda. In some cases, the arguments call only for *some* role; in others a *special* role; and in still others an *exclusive* role. In all cases, family farms are considered to be of value. "Family farms" is not defined at the outset to allow individuals who would defend the value or necessity of family farms an opportunity to establish criteria for deciding what is or is not classified as a family farm or family farm structure. In some cases, a definition of "family farms" as well as "family farm research" may be crucial for the defense; in other cases, it may not be.

A number of diverse philosophical arguments have been made in favor of the family farm as it is variously defined, and these arguments have vastly different implications for the critical appraisal of agricultural research. Recent literature reveals three broad categories in which arguments in favor of family farms exist: (1) a moral/cultural category, (2) a political-economic one, and (3) a category that can roughly be called "sustainability arguments." Although family farm advocates occasionally employ arguments from all three categories, the categories reflect divergent points of view. Indeed, each represents an appeal to a different source or reason for the value of family farms.

Two distinctions cut across the arguments and positions defending the place, role, or value of family farms in agricultural research policy and practice. The first is the distinction between *ideological* and *performance-based* appraisals or critiques. And the second is the distinction between *substantive* proposals for reform and *procedural* or *political* proposals. Briefly, ideological critiques link the illegitimacy of the agricultural research agenda to deep and permeating mistakes in the philosophy underlying the agenda. Performance-based critiques find no such deep and disturbing problems in the basic philosophy, but rather find problems and mistakes in the translation of that philosophy into the research agenda and specific research projects. Substantive proposals for reform call for reprioritizing or changing the goals and values of the research institution, whereas procedural proposals call for changes in the process or institutional mechanisms through which goals and implementation strategies are determined.

Before proceeding further, it is worth noting that some of the critical arguments now stated by family farm advocates are not new. Indeed, the nature, mission, and consequences of mainstream agricultural (and natural resources) research in the U.S. have long been subjects of concern and debate. As early as the 1870s, farmers resisted the "scientific agriculture" promoted by land-grant colleges (Kirkendall, 1986). In the 1880s and 1890s, southern populists argued that the land-grant system was part of a conspiracy launched in Washington to coopt and subdue their political and economic aspirations (Goodwyn, 1978). Social scientists in the early part of this century argued that the effects of agricultural research, especially technology development, would undermine the vitality of family farms and farm labor (Busch & Lacy, 1983). Social scientific analyses of the effects of research and technology on agriculture was rampant in the 1960s and 1970s. In fact, rising concerns about the effects of the research agenda during this era resulted in litigation. In 1978, the California Agrarian Action Project (CAAP) and California Rural Legal Assistance (CRLA) filed a lawsuit against the University of California (UC)-Davis, charging that UC's research on mechanical vegetable (tomato) harvesters resulted in deliberate harm to family farms and farm labor.

Despite the long tradition of critique and political challenge to the agricultural research establishment, the nature of the concerns and the tone of the debates have recently assumed a philosophical character. Earlier discussions cited and/or utilized agronomic, economic, and sociological data or case studies to argue the beneficial, harmful, legitimate, or illegitimate features of agricultural research. Now discussion focuses on the philosophies, values, politics, ethics, and, in some cases, epistemological commitments involved in such research. Social scientific analysis is still essential in understanding the nature, mission, and consequences of the agricultural research enterprise in relation to family farms and a host of other entities. But now philosophy seems to matter too.

Philosophy and the Family Farm

What, then, are the philosophical issues associated with the role and/or value of the family farm in agricultural research? There are five main philosophical issues:

- Of what value are family farms?
- What is the legitimate purpose of agricultural research, particularly publicly funded research?
- Should family farm values or interests be included in the research agenda?
- How should those values or interests be included?
- Who should answer each of the above questions?

These are clearly normative questions. Answers to them reflect basic values and/or ethical and political principles, and, in some cases, fundamental visions about how the world is and how it ought to be. Some answers may reflect strong feelings, and even deep religious beliefs and commitments, as Gary Comstock recently illustrated so insightfully (Comstock, 1987). Answers to others may reflect scientific and professional beliefs. One of the difficulties in philosophically resolving disagreements among various parties in the family farm debate is that many of these values and beliefs have *prima facie* legitimacy: Good reasons probably exist that support most of the positions advocated. The dilemma, or perhaps, the tragedy (Busch, 1989), is that the business of research requires that matters of ethics, values, political philosophy, and epistemology be settled, at least temporarily. As a consequence, some legitimate values and interests may not be considered, and therefore go unsatisfied.

This suggests that what may ultimately be at issue in the matter of the family farm and agricultural research is the nature of our political institutions. Perhaps we would not experience these tragedies if our institutions were structured differently, or guided by a different philosophy. Perhaps significant changes in these areas are not possible. Nevertheless, the family farm question leads to deeper questions about social ethics and the role of science in a democratic society.

One fact is clear: Farms are disappearing. And most of these lost farms were family-owned and operated. Research by social scientists has shown that farms are going out of business or being reduced to part-time status at a steady if not increasing rate (Figure 3.1). There are many reasons for the decline in the number of family farms—from the decision to enter another (perhaps more urban) occupation, to poor management and strong competition, to federal farm commodity policy, to international economics (Penn, 1979). Defenders of the family farm or a family farm structure, however, argue that the agricultural research system is the main contributor to the decrease in family farms. Whether intended or unintended, the research agenda has undermined the competitiveness and survival possibilities for smaller, more labor-intensive, family farming operations. According to critics, something is wrong with the philosophy of our research institutions, or with the interpretation and articulation of that philosophy in agricultural research practice.

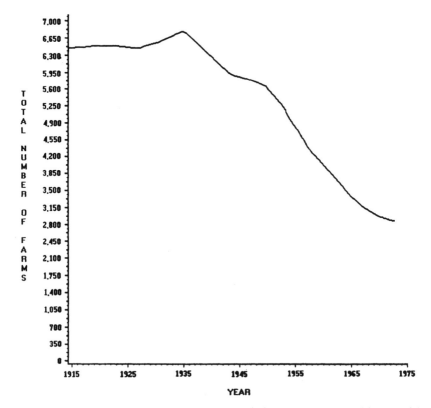

FIGURE 3.1. Total number of farms, United States, 1915-1973 (thousands). *Source:* 1915-1970, *Number of Farms, Land in Farms,* USDA, Statistical Reporting Service, Crop Reporting Board, Statistical Bulletins 316 and 507; 1971-1973, *Agricultural Statistics,* 1977, USDA.

Thus critical philosophizing must enter the picture. Is it true that research institutions have caused or will continue to cause the demise of the family farm? Is it true that something is wrong with this situation? If so, what? Why should agricultural research institutions modify their policies and procedures? Of what value are family farms anyway? First, we will address this last basic philosophical question.

Why Family Farms?

Popular and academic references to the farm crisis have been abundant throughout the 1970s, and especially in the 1980s. Major motion pictures, television news, magazines such as *Newsweek* and *Time*, an overabundance of scholarly books, and issues of journals, such as *Agriculture and Human Values*, have brought attention to the plight of the family farmer and the changing structure of agriculture.

What constitutes this crisis? Both sympathetic and neutral observers of the American farm situation appear to use the term "family farm crisis," or "farm crisis" to refer to one or more of a number of related situations.

A Personal or Individual Problem Situation

Farmers who are heavily in debt and unable to make payments on their loans because of crop losses, market conditions, or both may find themselves in a crisis situation. They may have to refinance the farm, take other employment, declare bankruptcy, or pursue other alternatives to protect themselves and their family. Sadly, a number of American farmers facing difficult financial times in the 1980s, found no option but suicide—or murder. These are the kinds of crises highlighted in the media—specific cases of threatened losses which repeat themselves across the nation. They are devastating for the individuals involved, but when most people allude to the family farm crisis they are probably referring to a more general and widespread occurrence of such situations.

A Class Crisis

Suppose a large number of farmers in the $40,000 to $100,000 annual revenue group, for example, come to face difficult or nearly impossible financial problems. Since family farmers are most likely to be in this middle income group, this situation can be described as a crisis for family farm as a class. However, in addition to personal and family crises, the situation extends beyond individual farmers and their families. Indeed, rural communities, migrant and nonmigrant farm laborers, urban consumers of certain commodities, and in fact, all who are negatively impacted by the aggregate loss of these farming operations are affected by the crisis. When an entire group of producers disappears, the rural communities that depended on them become victims as well (Goldschmidt, 1978).

The immediate social and economic effects that follow the loss or reduction in the number of family-operated, medium-income farms may not be that critical if other economic entities, employment opportunities, and commodity suppliers surface to fill the gap. In this case, the class-related or sector-wide crisis may simply be an "adjustment" problem, as some economists have maintained (Peterson, 1986). The family farm crisis may be no more significant than the loss of blacksmiths or mom-and-pop neighborhood grocery stores. These losses may be devastating for individual blacksmiths, grocers, and farmers, but they are inevitable results of progress. Recognizing this, some farm crisis analysts try to argue that the loss of medium-sized family farms result in impacts beyond the lost labor opportunities or commodity suppliers. Instead, the farm crisis entails a loss of a range of larger values.

A Social Crisis

The conclusion that the loss of family farms constitutes a larger social crisis—with the accompanying call for society, especially through government policy, to secure the family farm against internal and external threats—is based on arguments concerning the social value of these entities. All of these arguments recognize that the number of family farms is declining. However, three very different implications are drawn from this fact: (1) Our culture or moral tradition is therefore in decline; (2) the political and economic viability of our market-oriented democracy is in danger; and (3) the sustainability of the agricultural production system, and indeed the quality of life for present and future generations, is being jeopardized. There are overlaps, but for analysis, let us refer to these three as the *moral-cultural* position, the *political-economic* position, and the *sustainability* position, respectively. Each position includes an ethical or value-based assessment of family farms, and an assessment of our collective moral responsibilities in the current situation.

Because of the explicit reference to larger ethical or political values, this last definition of the farm crisis is the one that merits the closest philosophical attention.

Moral-Cultural Position

The term "agrarian" generally refers to philosophies favoring family farms. Agrarianism has a long history both in the U.S. and around the world (Montmarquet, 1989). One of its most recent proponents is Wendell Berry, whose views typify the *moral-cultural* variant on the agrarian theme. In *The Unsettling of America* (1977), and in subsequent publications, Berry argues that the farm crisis is a "crisis of culture," in which the demise of the family farm indicates the demise of both a cultural symbol and a set of moral ideals or virtues. The symbol is that of hardworking, self-reliant individuals participating actively and caringly in a community of such individuals. On one level, these individuals constitute a family community, and on a larger level, it is a village or small town rural community. Berry argues that the images of family farms and farm communities go straight to the heart of American culture and affect our thinking about ordinary morality. As such, family farms help form our concepts of responsibility to self and others, caring and sharing, and weathering adversity together. In Berry's view, these notions evolved in the context of early, farming-based, American culture and find their basic meaning in that culture.

The farther the image of the family farming community is removed from our national and collective consciousness, the farther we grow away from who we, as Americans, really are. The demise of the family farm is therefore part of a deepening alienation of Americans from what constitutes us as "a people" —our tradition, our land, our friends, and our families. On one level, Berry's assessment of the farm crisis is a broader philosophical indictment of "modernism" and self-serving individualism in contemporary America.

On a more concrete level, however, Berry's argument is that family farms demand and instill values and virtues that are morally and psychologically important for individuals and functionally necessary for societies. Among these are independence, self-reliance, circumspection, inquisitiveness, self-worth, craftsmanship, and generosity. Family farming requires at least the attempt to practice these virtues, and the "good farmer" reinforces these virtues or character traits in himself, his friends and neighbors, and especially his children. Berry's description of the good farmer tells us what he means by family farms: traditional, labor-intensive, and multi- and inter-cropped growing systems. Good farmers would disdain the modern conception of "production systems" as too closely aligned with modern notions of efficiency and productivity.

Berry calls for integration of crop and animal "growing," implying that the kinds of crop production that organic farming proponents advocate and the types of animal treatment that animal welfare activists promote are also a part of "good farming." Most importantly, however, is the idea that good farmers are more than efficient managers or users of scientific cropping and management techniques. Instead, they are caretakers and stewards of a morally sound way of life. This outlook on farming provides the foundation for Berry's assessment of agricultural research.

In Berry's opinion, the agricultural research system has deliberately undermined this way of life. The emphasis of the research agenda on efficiency, productivity, technology, and agriculture as a business has been, and continues to be, contrary to our moral values. In short, agricultural research is morally flawed. According to Berry, farmers (and everyone else) are morally obligated to reject modernism, with its emphasis on efficiency, productivity, and profits. We should reject both the products and the institutions of agricultural research unless or until the research system is totally redefined and assists family farmers exclusively. But Berry says this may not be possible. In any event, all farmers should cultivate their own characters and the characters of their children as diligently and carefully as Amish farmers cultivate their land and livestock. If they do, he says, they will be better, more whole people, and society will achieve a state of moral balance within itself and with nature. Berry's vision is indeed utopian, and is intended as an antidote to the mainstream vision of modernists in our society who see nothing in the farm crisis but reduced numbers, economic adjustment, or the inevitable result of the workings of capitalist economics.[1]

Political-Economic Position

In contrast to Berry's poetic vision, Marty Strange offers a "new economic vision" which is straightforward and realistic in his book, *Family Farming: A New Economic Vision* (1988). Strange's work typifies *political-economic agrarianism*. It is directed at political-economic policy with the goal of reforming policies so family farmers can survive. His argument is that the "farm crisis" is a financial crisis for family farmers in the sense of the "class crisis" described above, and a "social crisis" in the sense that our society politically and economically needs family farms.

Strange begins his analysis by attacking the notion that family farms are simply intermediate-sized farms. Family farms, he says, are a diverse lot. They differ in: (1) legal ownership patterns (incorporated, proprietorship, limited partnership); (2) acreage needs; (3) cropping and/or livestock operations; (4) capital requirements; and (5) management arrangements (e.g., individual, husband/wife, father/son, mother/daughter, etc.). Some family farms are traditional, and some are modern. Some family farms earn large profits, and some merely survive. The problem, according to Strange, is that an increasing number of family farms are neither profiting nor surviving. One reason is that in the current political economy of agriculture, farms have to "get big or get out."

According to Strange, it is not, at least in theory, economically necessary to either get big or get out of agriculture, contrary to popular wisdom and the views of many economists. Economies of size have been determined by a general economic climate promoted by the government since the 1960s and especially the early 1970s, when former President Nixon's secretary of agriculture, Earl Butz, urged farmers to take advantage of an expanded world market ("plant fencerow to fencerow"). Farmers responded by heavily mortgaging their land to purchase additional land and equipment. Financial institutions, including U.S. government agencies, were more than happy to underwrite this expansion. The market contractions of the 1980s left many farms, especially those trying to get big, "overleveraged" (with high debt-to-asset ratios) and facing high mortgage and interest payments on loans. Termed "debt obesity" by Strange, this situation has resulted in foreclosures on smaller farms. The foreclosed land and equipment were purchased by already large, "underleveraged" farms or nonfarm corporate investors, thereby further increasing the size and concentration of ownership of farming operations.

Farmers are in part to blame for the consolidation of agricultural production because they acted, at least initially, based on their desire to maximize production and profits. However, Strange says they had few alternatives. Invoking Willard Cochrane's (1979) notion of a "technology treadmill," Strange explains how farmers, in addition to acquiring land, had to adopt increasingly sophisticated and costly technology to remain competitive. This perpetuated the farmer's financial burden. Both loan payments and upkeep on the mechanical and chemical inputs now indispensable for production simply added additional overhead. In addition, former President Reagan's commodity policies fueled the high-tech and "bigger is better" trend in agriculture, allowing tax write-offs to purchase certain inputs. Most importantly, determining price supports by total production rather than productivity allowed some of the largest farms to reap the largest government payments which more than satisfied debt loads, thus allowing large farms to acquire financially troubled smaller farms.

In Strange's opinion this situation has been costly. The government's expenditure of commodity payments to financially secure and profitable farms has been passed along to taxpayers. The potential consumer costs of food produced by an increasingly monopolistic production system cannot be accurately estimated. However, Strange says if the trend continues, even with some limits on the incentive to expand that are embodied in the 1986 Tax Reform Act, we will face escalating food prices. Moreover, social and political consequences are associated with medium-sized family farms. The first is the loss of freedom of choice. As long as family farmers must: (1) farm more land than they can individually farm to survive; (2) borrow excessively for land and equipment; and (3) become a part of the technology treadmill, their freedom is restricted. What if an individual farm family wants to neither get big nor get out? The second consequence is the loss of equality of opportunity. Without a stock of capital sufficient to underwrite the land and technology requirements for a large farm, an individual cannot enter farming as an occupation.

Most importantly, the political and economic power that can be wielded by individuals in an increasingly concentrated agricultural production system threatens to undermine both the democratic process and the freedom of free markets. In recent years, political scientists have noted how the once-powerful "farm bloc" consisting of the United States Department of Agriculture (USDA), the Farm Bureau, and congressional representatives from farm states now are being overpowered by a diverse and internally embattled collection of commodity and special interest groups, including environmental advocates and labor interests. None of these diverse groups are uniquely positioned to influence public policy regarding agriculture (Browne, 1988). However, if a new farm bloc should arise—this time consisting of a very few large agribusiness conglomerates and allied large corporate farms—the financial power and control of resources this bloc maintains could undermine our society's ability to socially control agriculture, either through government policy or market power.

Strange's new economic vision is that a farm structure composed of family farms can perform as well, and probably better than an industrialized and concentrated agricultural production system. Family farms can produce enough food for a predominately urban population as efficiently and inexpensively as large-scale corporate farms. Family farms can utilize resources as or more efficiently and wisely than agribusiness conglomerates. However, high-cost, resource-depleting technologies continue to be foisted upon farmers by agricultural researchers and agribusiness salespeople.

According to Strange, farmers (and everyone else) could do fine without center-pivot irrigation systems, large swine-growing buildings ("pork palaces"), and other costly inputs such as bovine somatotropin ("bovine growth hormone"). If the government changed its tax and credit policies, family farms might have a chance, and if agricultural research would stop developing and promoting only large-scale technologies, more family farms could survive. Then, if more family farms survive, stagnant and even dying rural communities could be revitalized.

The fundamental idea of Strange's philosophy is that the opportunity to enter farming or to remain in farming must be preserved, or one more freedom will be taken away in the hasty march of progress. Strange concludes that we must return political, economic, and farming power to the people.

Sustainability Position

The third category of arguments we will consider concerns the role of family farms in achieving sustainability. Some distinctions are appropriate here, because not all advocates of sustainable agriculture believe family farms have value. While all maintain that "sustainability" is a good, or even morally obligatory state to achieve, some advocates of sustainable agriculture believe that any new technology developments will contribute to sustainability. Others call for caution in resource use, regardless of the size or type of farming operation. And some advocates maintain that only through traditional, small-scale farming can agriculture be sustained (Douglass, 1984). This last position has clear similarities with Berry's position, though the basic reason to favor family farms differs.

The position we will focus on now differs from all of the above, however. This position emphasizes the value of family farms in achieving sustainability, but not their exclusive value in this regard. That is, the value of family farms does not have to do with family farms *per se*, but with any institution that contributes to securing and sustaining a diverse set of values, not the least of which is enough food and a decent quality of life for future generations of human beings (Burkhardt, 1989).

According to this last sustainability argument, family farms have value if they (or a structure containing them) contributes to and/or does not undermine our attempts at sustaining food production. However, an additional argument is made that they have value because they can contribute to sustaining basic human values, such as fairness and respect for future generations and the environment, rather than because they simply contribute to sustaining agricultural production. Accordingly, to the extent that a family farm does not contribute, or perhaps undermines these values, it cannot be considered valuable. This perspective fragments the concept of "family farms" even more than Strange's. There are large, small, full-time, and part-time family farms. There are productive ones and unproductive ones. There are also racist family farms, as well as sexist, militant nationalistic, environmentally sound, selfish, and community-destructive family farms. Some family farms practice soil and water conservation and some do not. There is thus no necessary connection between important human values and farm size, ownership, or manner of operation.

Sustainability proponents do, however, find two contingent but probable connections between family farms and important social and environmental values. The first is that family-owned and operated farms, with individuals aware of and concerned with environmental soundness and social (and personal) justice are more likely to translate values into on-farm and off-farm practices than corporately owned farms operated bureaucratically—like other large businesses (Burkhardt, 1986). Much depends on the individuals involved, but family farms, because of their size, may simply have to practice more conserving and diversified farming practices. Greater managerial efficiency, environmental concern, and positive impacts on rural communities may be the result.

The second connection between family farms and broader social values is that an agricultural structure containing family farms may help retain an element of cultural, biological, and economic diversity. Diversity is considered to be good, not for its own sake, but because it contributes to sustainability—in essence, an "insurance policy" against potential problems associated with the environment and political-economic institutions. Note that diversity itself includes the value of not only family farms, but also a wide spectrum of farm size, ownership patterns, and production strategies.

Thus, not all family farms have value. Their value is subject to the determination that each farm practices sustainable agricultural practices and upholds other basic values such as community and justice. A diverse set of individuals acting on and promoting these values is preferable to all family farms or all large-scale corporate farms. Because of the trend toward homogeneous, large-scale corporate agriculture, however, the sustainability position ultimately calls for the protection of family farms. Family farms can also respect important social values and should be secured against cultural, political, and economic decline.

The sustainability position thus acknowledges the philosophical merit of moral-cultural and political-economic agrarianism. Traditional family farming is a good thing—among many good things—and family farm interests need to be protected and promoted. According to the sustainability position, however, family farms, along with all other kinds of farming operations, still must be evaluated based on their relative contribution to the value of sustainability.

Other arguments have been promoted regarding the value of the family farm. As noted, Gary Comstock (1987) has identified religious reasons (some akin to Berry's), as well as aesthetic and nostalgia reasons, among others. I have argued (Burkhardt, 1989) that the moral-cultural, political-economic, and sustainability positions are more philosophically defensible, than these other reasons, and that sustainability arguments are the most convincing. But, let us address the assessments of the agricultural research system's agenda that relate to each of these perspectives.

Table 3.1. USDA Goals of Agricultural Research

I. Insure a stable and productive agriculture for the future through wise management of natural resources.

II. Protect forests, crops, and livestock from insects, diseases, and other hazards.

III. Produce an adequate supply of farm and forest products at decreasing real production costs.

IV. Expand the demand for farm and forest products by developing new and improved products and processes and by enhancing product quality.

V. Improve efficiency in the marketing system.

VI. Expand export markets and assist developing nations.

VII. Protect consumer health and improve nutrition and well-being of the American people.

VIII. Assist rural Americans to improve their level of living.

IX. Promote community improvement including development of beauty, recreation, environment, economic opportunity, and public services.

X. Enhance the national capacity to develop and disseminate new knowledge and new or improved methodology for solving current problems or new problems that will arise in the future. Research under Goal X is conducted under all the RPAs [Research Problem Areas] of Goals I - IX.

Source: Busch & Lacy (1983). Reprinted by permission.

The Agricultural Research Agenda: Productionism and Its Critics

In the U.S., a "settled public philosophy" seems to exist about family farms. This philosophy is institutionalized in larger policy and in agricultural research policy and practice. The USDA has goals for agricultural research (Table 3.1), and helping family farmers is not explicitly among them. If these goals influence research priorities for the Agricultural Research Service (ARS), the land-grant university system, and state agricultural experiment stations, then family farm research is not a high priority. Many administrators, scientists, and observers believe that most research helps family farms (see, e.g., Tweeten, 1987), and until recently, most scientists and administrators had farm backgrounds (Busch & Lacy, 1983). Still, a minimal amount of explicitly family farm research has been conducted. Perhaps this is because of value-free scientific pursuits. Perhaps it is because family farm research is not an official goal, even though "helping family farmers" is routinely mentioned in administrators' pronouncements and political rhetoric. Many goals of the USDA, if acted upon, may help family farms (for example, by serving rural communities). But in the opinion of some critics, the failure to deliberately help family farms is deliberately harming them.

Agricultural research policy in the U.S. that is institutionalized in the USDA/land-grant system formalizes and articulates the underlying philosophy or purpose of the research enterprise. The philosophy was "to serve agriculture," and the mission was to "rationalize" agriculture. Rationalizing agriculture meant introducing scientific concepts and practices and the most sophisticated "technical systems" available for the production of food and fiber. From an historical perspective, the reasons behind the government's desire for rationalization in the 1860s seem reasonable:

- Continuing development of an industrial nation depended on an increasingly urban-centered population. Growing urban centers and increased population meant greater total production was needed from each farm.
- International balance-of-payments problems created by war-debt financing demanded greater agricultural output for exports.
- The abolition of slavery and the breakup of the plantation system meant a decrease in farm labor (or increase in labor costs), demanding the substitution of capital-intensive for labor-intensive agriculture.
- Advances in the biological and engineering sciences—and the development of "scientific management"—allowed significant increases in production and productivity (Busch & Lacy, 1983).

From its inception, then, the land-grant system promoted "productionism" (Shepard & Hamlin, in production), closely allied with "scientism" (Rosenberg, 1976). Advocates of the new scientific agriculture promoted new management techniques, technologies, and distribution networks. As noted earlier, farmers frequently resisted these new tools, especially in the South and upper Midwest. Many were undoubtedly afraid of the "gadgets" and "eggheads" from the land-grant universities. Many farmers predicted early that a technology treadmill was the likely outcome of scientific agriculture (Cochrane, 1979). Many saw the "rationalization of agriculture" in political terms— service to Northeastern financial and industrial interests (Goodwyn, 1978). Nevertheless, as Ruttan (1982) has noted, the scarcity of labor and the abundance of arable land in the U.S. dictated that science and technology-based agriculture would emerge—and even dominate farming.

Although questions have been raised about the effects of rationalization since the inception of the land-grant universities, productionism, the rationalization of agriculture through science and technology, and the emphasis on efficiency have continued to permeate the public agricultural agenda. The official assessment of the results of this philosophy and mission is that American consumers benefit from having the lowest food prices-to-income ratio in the world (Tweeten, 1987). American farmers are considered to enjoy a standard of living well above that of farmers anywhere, and not too different from American urbanites (Easterbrook, 1985). The new technologies, including mechanical harvesters; agrichemicals; and, most recently, biotechnologically engineered plant varieties; computer-driven irrigation systems; and sophisticated post-harvest transport, wholesaling, and retailing systems are proclaimed to have allowed agriculturalists to all but overcome the inherent risks involved in farming (Penn, 1979).

There is little doubt that the infusion of U.S.-developed technologies into less-developed nations, and the ensuing "Green Revolution," which transformed not only agriculture but whole societies, have raised the standard of living for millions of people (Griffin, 1972). Clearly, numerous beneficial aspects have resulted from the productionist philosophy and the rationalization of agriculture through agricultural research, education, and extension. For most agricultural researchers and administrators, changing either the philosophy or the agenda is not necessary, except perhaps to make it even more scientific.

According to some agrarians, however, the research agenda is morally flawed, despite the benefits it has generated. Berry nearly goes as far as the early populists to claim that the "inevitable" result of the rationalization of agriculture—the demise of family farms—was and is intentional, and is all the more morally to blame. He argues that scientific agriculture is based on "big science." Scientists and administrators prefer, for both scientific and political reasons, large, heavily funded research projects. Scientists and administrators like big machines such as 400-horsepower harvesters and high-power computer systems. These projects intentionally favor large, profitable consumers of the technology—large farms, agribusiness corporations, and the military-industrial complex. Berry rejects the philosophy of productionism, the idea of rationalization, and such notions of scientific progress and efficiency. (Traditional) family farms do not need and are undermined by the agricultural research establishment. And the agricultural research establishment does not need—nor want—traditional family farms. According to Berry, this is wrong. The basic philosophy of the institution contradicts the basic moral values and cultural traditions it was commissioned to serve. The agricultural research system should therefore either totally redefine its philosophy and mission, or shut down. Berry's position is an ideology critique with substantive demands for change.

According to Strange, there is no logical connection between scientific agriculture and the demise of family farms. Scientific agriculture may be capable of serving the interests and needs of family farmers while also serving the needs of urban consumers. Efficient, high-yielding agriculture can be accomplished with appropriate technologies; appropriate, that is, to family farm operations. However, to accomplish these objectives, research design and technology development must be reconceptualized: Family farm research must be among the high priority goals of agricultural research. The fact that it is not a high priority suggests that decision-makers have either been blind, or blinded by other interests and goals.

For some agrarians, the negative effects of agricultural research on family farms may simply be the *unintended consequences* of productionism. Productionism, together with the deliberate strategies of large-scale agribusiness corporations to direct the research effort toward their goals, has caused small farms to be squeezed out. Underlying this view is the notion that, in any institution, "philosophical inertia" can result. Productionism and rationalization strategies were originally intended to help family farmers—the only farmers that existed in the late 19th century. But as family farmers became successful, and as other interests, including nonfarm corporate interests, became involved in agriculture, that philosophy and mission inadvertently became a disservice to family farms. According to this interpretation, agricultural research now benefits large farms and agribusiness firms, and indirectly harms small family farms because the philosophy and mission of research have not changed. This is not so much an ideology critique as a performance critique, but with a substantive proposal for revision.

A somewhat more critical position is taken by Hightower (1973) and others, such as Kenney (1986) and Friedland et al. (1981). Their position is that the philosophy has indeed changed: "Serving agriculture" and productionism have been replaced by "serving corporate America"—nonfarm agribusiness interests—and deliberate attempts are being made to tie farmers with all sizes and kinds of operations to multinational industrial and financial corporations. Science and technology development performed within the land-grant universities and USDA are now so dependent on and aligned with big business, nothing short of a major political-economic transformation and an accompanying scientific revolution (Kuhn, 1970) within the research system will return agricultural research to its original purpose: serving family farms. This view appears to embody an ideology critique directed toward reforming the institution's philosophy. However, whether a substantive revision of the institution is needed or simply more democratic participation in the agenda setting process is unclear. In either event, however, moral revision is demanded.

The evaluation of agricultural research from the sustainability perspective can take a number of directions. One approach says that because science and productionism are not essentially contradictory to sustainability, our concern would be to institute more sustainability-oriented research within the current system. According to this perspective, ecologically sound technologies and resource-conserving management systems are not only not incompatible with science, but in fact need a scientific base (Lockeretz, 1983). Thus, rather than change the philosophy and structure of the institutions, we should establish and fund projects which are simply more sustainable—including projects such as the Low-Input Sustainable Agriculture (LISA) program (Madden, 1989). The indirect beneficiary of such research will be family farmers. This is a performance-based approach, with some request for substantive change, at least in agenda priorities.

For other advocates of sustainability, however, the goal is not simply to conduct more or better research on the technical dimensions of sustainability. Because not enough research has been performed on the community or human aspects of sustainability, including the place of family farms in a sustainable agricultural system, something is wrong with the philosophy of productionism and concomitant scientism. "Techophilia" (Drengson, 1982) in the agricultural research establishment precludes viewing research as ultimately aimed at solving human problems. Even farming systems research (Hildebrand, 1980), while ostensibly targeting small farms, has still failed to define farming systems in larger community and environmental systems terms (Oasa & Swanson, 1985).

Despite the documented concern for rural life in the USDA's goals, the result of the research agenda has been that new and more technologies have been introduced into agriculture, and family farmers have gone out of business. Consequently, rural communities have suffered and some have ceased to exist. Researchers and administrators have misunderstood the reasons why public institutions should serve agriculture: to serve "the public." Continuing to increase production in an era when many crops have yearly surpluses and farms and towns are dying is a fundamental misunderstanding of our public philosophy. The long-term consequences of this mistranslation on family farms and on the sustainability of the system demand change. This

position constitutes an ideology critique in which substantive and procedural revisions in the agenda-setting process are also demanded.

Moral Dimensions of Research Policy

A variety of conclusions can be reached from the critiques discussed
in relation to the moral legitimacy or illegitimacy of the research agenda.
In a basic philosophical sense, there are two sets of arguments. In the
first set, family farms are considered valuable for one of three main
reasons: (1) family farms embody and transmit basic moral values; (2)
family farms are socially important economic and political entities; or
(3) family farms, when "morally managed," can contribute like no other
socio-economic entity to sustain important environmental and social
values. Policies and practices that harm family farms are considered
wrong for one of these reasons.

In the second set of arguments, the agricultural research system is
morally evaluated. Two kinds of arguments are included in this set:

1. Arguments which blame the agricultural research establishment
   for harming family farms. Either the research establishment was
   intentionally designed to promote scientific and technological
   changes inimical to family farming operations, or these results
   were only accidental. In both cases, however, individuals in the
   research decision-making process acted (and continue to act) im-
   morally.
2. Arguments which attribute no blame but do propose changes.
   Regardless of the historical harm to family farms, the agricultural
   research system can and should produce technologies and research products which benefit family farms.

Let us assume for the moment that harming family farms is *prima facie*
wrong. Consider then Berry's position that agricultural research explicitly
and intentionally promoted the demise of small, traditional farms. In a
radical interpretation of this view, decisions made by the agricultural
research establishment have the moral characteristics of Hitler's "final
solution"—a reprehensible systematic attempt to exterminate a class of
individuals. Even though this interpretation may be rhetorically useful
for anti-research establishment "believers," a more plausible interpretation is that the action reflected a perceived trade-off between family
farms and the need for an increasingly efficient and productive agricultural production system. As noted previously, early observers of
agricultural research and technology development did predict that some
farmers would go out of business as a result of the system, and some

establishment decision-makers possibly were aware of these predictions.

Hadwiger (1982) claims that the entire agricultural scientific enterprise has been and continues to be "utilitarian," by which he means concerned with producing "useful results." But more is included in the implicit utilitarianism of agricultural research policy-making. Policy-makers at higher levels of power and authority generally tend to take a "god's-eye" view regarding the impacts and consequences of their policies. They aim to achieve public interests, which tend to be interpreted in terms of trade-offs or "balancing of interests" (Benn & Peters, 1959). As such, an action perceived to be beneficial to a large population of urbanites, although at the expense of harming a smaller number of rural inhabitants, would generally be undertaken. Policy-makers may have determined that the demise of the family farm was worth the risk given the relatively minimal social benefits family farmers provided.

It is more likely, however, that when publicly funded agricultural research was established, the decision was made not to destroy family farms, but to promote rationalization and productionism, wherever it would work. The trade-off was not between family farms and production, but between less production and more production. The demise of small farms, although possible, was not intentional and maybe even contrary to the intentions of decision-makers.

Even if this last scenario models historical reality, a number of negative evaluations might follow (assuming again that family farms have value). One evaluation is based on a moral equivalent of the legal doctrine of "strict liability." According to this doctrine, if the harm is great enough, we are justified in attributing blame even though the agents involved neither knew about nor intended to cause that harm. The argument is that since the harm to family farms has been excessive, the research establishment is morally at fault. Indeed, agricultural research decision-makers should have known that their decisions would harm family farmers.

The difficulty with strict liability attributions is proving both the "degree of harm" and "should have known" components of the assessment. In the case of harm to family farms, both are problematical. As indicated, unless family farms have intrinsic value, substituting them for other values, intentionally or unintentionally, may not be that morally problematical if other values are more valuable. Moreover, if research establishment decision-makers actually intended to help family farmers, attributions of moral blame seem inappropriate, and perhaps morally wrong in and of themselves. Perhaps decision-makers made mistakes, but without fault. The grounds are therefore weak for morally blaming past decision-makers for their intentions and/or the unintended consequences of their actions regarding family farms. However, the actions of current decision-makers have affected and could continue to negatively affect family farms. Should we hold them morally accountable and to blame?

Moral blameworthiness is dependent on a number of factors: whether the action is wrong, whether the agent knows it is wrong, and whether the agent intends to do that wrong. As we have stated, harming family farms is *prima facie* wrong. One question is whether or not present scientists and administrators know or believe that it is wrong to harm family farmers. Most of these individuals probably believe it is wrong to harm anyone, but believe one of three things: (1) the research agenda does not harm and, in fact, helps family farms (Tweeten, 1987); (2) the research agenda itself has no social consequences, but has consequences dependent on larger government policy and the operation of the market; and (3) some utilitarian criterion continues to guide our decision-making, and implies that while harm to family farms is *prima facie* wrong, it is not wrong when all things are considered.

It should be clear that (1) and (2) reflect differing beliefs, and perhaps even theories, about the relationship between the research agenda and its social impacts. Blaming individuals for their beliefs and theories is difficult and perhaps morally impossible unless a strict liability argument is made. As illustrated with past decision-makers, strict liability assignments are inappropriate in this case. The question of blame really focuses on the legitimacy of using the utilitarian calculus or some variety of it. Is it wrong to do this?

Agrarians often question the specific utilitarian criterion used in the agenda-setting process. Strange appears to have this in mind when he argues that decision-makers are at fault for interpreting costs and benefits in a strictly economic sense. That is, they fail to see that more significant social, political, and perhaps environmental costs are associated with particular research and development agenda items. Failing to consider these means failing to include family farms in the research agenda. If these other costs associated with the loss of family farms were "calculated in," the resulting cost/benefit analysis would indicate the value of family farm research. Family farms may have been unintentionally harmed in the past, but with modifications in the decision criterion, the research agenda can now benefit the larger society, while also benefiting family farms.

A stronger criticism rejects the utilitarian model altogether. According to this view, some values, including "basic values" cannot be measured, and thus cannot be calculated into utilitarian, cost-benefit assessments. According to this perspective, values such as human health, fairness, and environmental quality are values a society must simply promote and protect, independent of any trade-off analysis (Sagoff, 1988; Rawls, 1971). Not harming family farmers is simply a matter of fairness (Comstock, 1988). Therefore, decision-makers should be blamed for not including, and even excluding, considerations such as fairness and effects on family farms in the research agenda.

There are two aspects of this criticism which are relevant. The first is the contention that administrators and scientists are utilitarians in some sense. This is debatable. As Busch and Lacy (1983) have shown, research agenda decisions tend to be made in accordance with canons of scientific curiosity, scientific adequacy, USDA goals, or simply to solve a problem as it arises. Therefore, to blame decision-makers for their unconscious utilitarianism seems inappropriate, and again, perhaps immoral in its own right.

The second aspect of the attribution of moral fault is that decision-makers fail to include certain values in their evaluation of research (utilitarianism aside). Perhaps they do not explicitly consider some values, the value of family farms in particular. However, to blame this and the effect of agricultural research on family farms on decision-makers from the highest levels of USDA to the bench science level may miss the point. If a moral problem exists, it is perhaps more with the "system" through which research is directed and funded than with the decision-makers themselves. The system may not reflect the deeper and larger political and social-ethical values we hold and may even preclude their consideration.

Busch and Lacy (1983) and Busch et al. (1990) present a model which attempts to capture the dynamics of the "demand for and supply of" agricultural research and technology. Contrary to critical analyses which attempt to lay blame on particular individuals in the research system, Busch, et al. argue that agricultural researchers and administrators are involved in a system of "pushes and pulls." This system effectively precludes the consideration of some values and social consequences in agenda-setting and research project choices. According to the push-pull model, the products of agricultural research (seeds, animal feeds, machines, chemicals, and management techniques) that make their way into agricultural production are generally the result of demands made on the research establishment at particular places and points in time.

However, these products are not supplied simply because they are "demanded by farmers," nor are they induced "from the outside" by researchers, USDA decision-makers, or corporate salespeople. Instead, a "dialectic" occurs between the research establishment and its clients. Each individual, whether individual or corporate, brings to the "market for research products" certain values, interests, needs, etc. Products which are ultimately "bought" or "sold" (originally metaphorically, but increasingly literally) are the result of negotiations among the individuals involved.

As illustrated in Figure 3.2, farmers bring to the market needs for survival, productivity, etc. This translates into demands for specific products, such as hybrid seed, and specific technologies, such as low-energy corn dryers. Researchers bring to the market the desire to use the most recent applicable scientific discoveries, which frequently translate into the most sophisticated machinery and (recently) biotechnologically generated seeds.

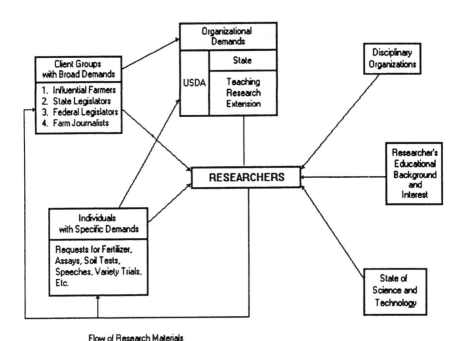

FIGURE 3.2 The push-pull model of agricultural research. *Source:* Busch et al. (1990). Reprinted by permission.

In the "negotiating process," scientists deliver "scientific translations" of what clients seem to be demanding. These translations are increasingly scientifically (and technologically) sophisticated, because, as Berry correctly states, scientists are scientists, and want to utilize the best scientific results in technology development. Similarly, clients (farmers, agribusiness firms, the military) want something from the research establishment. What they get may not be exactly what they wanted, but they will not use the research product if it is contrary to their interests.

The push-pull model helps explain both past and present agenda-setting for agricultural research, and shows why it is inappropriate to morally fault individuals within the research system for the negative effects of research or for failing to consider a range of values. Historically, the government's desire to make agriculture more productive influenced the structure of the research institution. The research projects scientists became involved in were influenced by that structure. The state of scientific knowledge also influenced decisions. Farmers' interests were considered as well, including their desire for increased profits. The confluence of "science" and "profits" led to the supply of increasingly scientifically based inputs—tractors, chemicals, hybrid seed, harvesters, and post-harvest technology. In theory, at least, the resulting rationalized agriculture was determined by "settled negotiations" between "farmers" and "researchers" (Busch, 1980).

This system favored and continues to favor larger farms, as well as agribusiness firms, consumer interests, multinational chemical corporations, and biotechnology firms. The reason is simple: Larger, successful farmers and businesses tend to use the agricultural research system. They tend to contact the research establishment more often than smaller farmers and they fund research and technology development. Individuals who contact administrators and scientists are frequently their former students and colleagues. The point is that a clearly identifiable group of individuals exert pushes and pulls on the system, and family farmers are not primary among them. Benefits clearly have accrued to non-family farming interests, and burdens have been placed on smaller family farms. The issue is whether a way to introduce or reaffirm family farm values into this push-pull system can be considered. Failure to consider these values once they are "inside" the system would indeed be grounds for moral fault.

Social Ethics of Agricultural Research

We know that new technologies and products based on biotechnology and computer technology, as well as those derived from more "traditional" biological and engineering sources, will continue to find their way into agriculture. And everyone, from research administrator up to national policy-maker, knows that farmers will continue to be affected. USDA and the Office of Technology Assessment (OTA) have published reports highlighting the expected continuing decline in the number of farms and increases in the size of farms that remain. Once again, the notion of "structural transformation" emerges as if this "march of modernism" is inevitable.

But the march is not inevitable. Such trends are the result of individual and social decisions and these decisions reflect values and principles. We indeed have reasons to find value in family farms. Deciding not to alter the institutional framework within which public decisions are made could be construed as a violation of the basic social-ethical principles to which we are committed. Among these principles is fairness.

Philosopher John Rawls has argued that, in a just society, each person is entitled to the most liberty compatible with like liberties for others. He has also argued that inequities are justifiable, but only if they are to the advantage of the least advantaged. Rawls' *A Theory of Justice* (1971) is not without its philosophical difficulties and critics, but one thing it does is explain what many consider the underlying philosophy of social ethics governing the U.S. constitution and the historical construction of our main social policies (Richards, 1977). It explains how individuals are entitled to "basic goods," among them equal respect and equal access to economic and political opportunities. Importantly, decisions about "public goods"—goods shared and financially supported by "the public"—must be public decisions. The public includes many individuals with diverse interests and values, and family farmers are part of the public.

The California Rural Legal Assistance Project's (CRLA) lawsuit against the University of California (UC)-Davis contends that the university conspired with agribusiness interests in California to develop the mechanical tomato harvester technological package. The package was designed to benefit large producers (and processors), and, according to CRLA, intended to reduce the need for farm labor as well as to drive smaller operators out of the market. Regardless of the intent in this case, the philosophical basis for the suit is roughly Rawlsian: Public goods demand public access as a matter of justice. The research performed on the tomato harvester differentially benefited large producers while in no way benefiting those disadvantaged by the diffusion of the resulting technology. The situation was not fair, just as many of the other research and technology development projects now underway are not fair (Comstock, 1988).

Lawsuits aside, a philosophical case can be made for the social value and justice of providing family farmers access to the research agenda decision-making process. The argument is straightforward: Family farms do have social value. And family farmers as a class have legitimate interests regarding agricultural research. The public institution either has intentionally, or more likely, unintentionally discriminated against family farm interests. Discrimination against legitimate interests is wrong as a matter of social justice (Blackstone & Heslep, 1970).

California's court-mandated "solution" to the CRLA lawsuit is that non-scientific and non-agribusiness interests must have a voice in reviewing proposed research at the public institution. In effect, these "new agenda" interests must now become part of the push-pull system. Representatives of urban, labor, and small farm interests are able to review and make recommendations regarding particular projects. This compromise situation is not without its critics: Scientists claim it slows the implementation of a research design, and "external" critics claim that the structure of the committee allows non-science and non-agribusiness interests only a token voice in the decision-making process. Those problems aside, the *idea* of other interests or values having representation in reviewing research projects, or even in agenda-setting, has considerable philosophical merit in protecting or respecting family farms in the future research agenda. Indeed, now that they are politically as well as morally considerable, failure to consider them in establishing the research agenda may once again be morally blameworthy.

Nevertheless, two problems are apparent with this approach. First, as stated above in the analysis of sustainability arguments, just because an individual is a family farmer does not mean that he or she is morally praiseworthy. Indeed, an individual family farmer may have values and interests that are inimical to the social values associated with the existence of family farms. For an individual family farmer to be a part of the research agenda-setting process does not guarantee that sustainable, just, and ecologically sound farming research will be conducted or even be considered in the institution. The values associated with family farms, rather than family farmers themselves, are what should be represented in the agenda-setting process and in subsequent research.

The second problem is that allowing these "external" individuals— non-science and non-agribusiness people—to participate in the decision-making process may have the unintended effect of permitting them to remain "external." The danger is that the values or interests they represent always will be residual to the main business of research— responding to pushes and pulls, conducting production-oriented research, etc. However, perhaps this is acceptable. As the sustainability position suggests, family farm research should be part of the agenda, not constitutive of the agenda. External participants may pull decision-makers in a slightly different direction, but not be capable of pushing their own agenda completely through. To not consider family farm values may be wrong, but to consider them while excluding other values may be equally wrong.

Conclusion

The idea that family farm values are socially important, and should therefore have a place in the agricultural research institution, ultimately rests on the convergence of two basic principles. The first is that public institutions are legitimate only if they serve the public—the principle of institutional legitimacy (Benn & Peters, 1959). The second principle is that because human activities are by nature purposive, science as a human activity is therefore necessarily oriented toward some purpose—the principle of intentionality (Habermas, 1972). The convergence of these two principles leads to the normative conclusion that public science must serve the public.

One difficulty with the principle of institutional legitimacy is that most contemporary societies are pluralistic: "The public" is, in fact, distinct publics. These publics frequently have conflicting values and interests, so service to one entails ignoring or even disserving others. This is an ancient problem in political philosophy—how to establish a "public interest" which satisfies each separate public, while being substantively meaningful in order to create law and policy. As noted, utilitarian strategies attempt to define the "greater good" in terms of trade-offs or balancing of interests. Non-utilitarian views appeal to our values, "the community," and, in some cases, basic principles of justice or morality. Convincing philosophical reasons exist for both positions, and framers of constitutions frequently attempt to employ both criteria in attempting to define the public good. Occasionally, these views come into conflict, and decision-makers simply must decide between them.

Nevertheless, the important point is that the principle of institutional legitimacy demands we attempt to find the public good, apart from the particular values and interests of the various publics competing for favor and even dominance in a society. In other words, the idea of "the public good," though ill-defined and imprecisely captured in both utilitarian and non-utilitarian philosophies, remains an ideal that individuals and institutions continue to strive to attain. Attaining it may not ever be possible, but is not inconceivable.

The intentionality of public science becomes clear in this regard. Science, in theory defined as the systematic pursuit of knowledge, is a human activity. Public science, that is, research and technology development funded through public sources, must ultimately be directed toward human needs. Those needs include knowledge for its own sake, but also the amelioration of the human condition. However, not every experiment or research program must be designed to explicitly and specifically benefit someone. Basic research, to the extent that categories such as "basic," "applied," and "development" even apply, may not serve a definable public goal other than providing more knowledge. Nevertheless, the ideal is that science generally will serve human needs, including the needs of far-distant future generations. This conclusion does not contradict the ideal of value-free science, as long as value freedom is understood to mean objectivity or a lack of bias (Rudner, 1972). Indeed, the meaning of objectivity is that instead of being directed toward particular interests or reflecting particular biases, science is directed toward the greater good.

Within public, institutionalized research in the U.S., we are as far from realizing both ideals as we are of realizing their convergence. Nevertheless, if we can assent to these ideals, we may be closer to realizing them. Indeed, because the discussion of science and agricultural research has assumed a philosophical tone, as alluded to in the beginning of this chapter, we may have already begun to assent to them. For agricultural researchers, administrators, and producers to be willing to discuss more than the technical or economic effects of research, is itself indicative of a fundamental shift in perception. Now these individuals are considering the values governing research—productionist values, family farm values, environmental values, nutritional values, justice, sustainability, etc. Each of these values, "new agenda items," or "external interests" will ultimately stand or fall on the basis of their philosophical merit and the arguments in support of their importance. Nevertheless, because they are included within the assessment of science suggests that progress is underway not only in science and technology development, but in social ethics as well.

Notes

1. In the theoretical literature on socialism in the early 20th century, a great debate took place about which should come first—a socialist government promoting cooperatives, communes, and social ownership of productive entities; or "socialist man," the individual able and willing to be a part of a communal enterprise—the unselfish individual. If an analogous debate exists within modern agrarianism, Wendell Berry would, in all probability, side with the latter. If we change ourselves, the system will change to accommodate us because we are the system.

4

Moral Responsibility in Agricultural Research

Vernon W. Ruttan

Introduction

In the 1960s and 1970s, a new skepticism emerged about the benefits of advances in science and technology (Shils, 1974). This view charged that the potential power created by the fusion of science and technology, as reflected in the cataclysm of war, the degradation of the environment, and the psychological cost of social change, is obviously dangerous to the modern world and to the future of humanity. As a result, the significance of scientific progress for human welfare, technical change, and economic growth was seriously questioned. After a decade of slow economic growth the extent of the concern has dissipated. But concerns about the ethical implications of advances in science and technology have intensified. (Ruttan, 1982a,b; 1983a,b).[1]

Agricultural science has not escaped these questions.[1] This chapter attempts to assist agricultural scientists and science administrators engaging in fruitful dialogue about the concerns of traditional rural clientele, advocates of a broader agenda for agricultural research, and populist critics of the agricultural research agenda that have affected agricultural research institutions. Before discussing some of the moral issues raised by agricultural research, however, some perspective is needed on the historical forces that have influenced the rate and direction of technical and institutional change in agriculture throughout the last century.

Sources of Technical Change

We are, in the closing decade of the 20th century, completing one of the most dramatic transitions in the history of agriculture. Prior to the beginning of this century, almost all increases in agricultural production occurred as a result of increases in area cultivated. The major exceptions were Western Europe, where livestock-based conservation systems of farming had developed, and East Asia, where wet rice cultivation systems had evolved. By the end of this century, few significant areas will remain where agricultural production can be expanded by simply adding more land to production. Expansion of agricultural output must be obtained almost entirely by more intensive

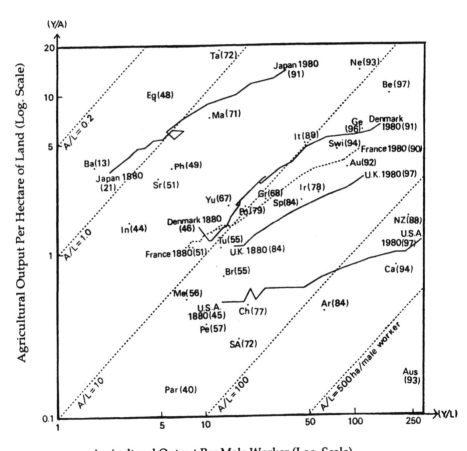

FIGURE 4.1 Long-term patterns of growth in output per hectare and output per worker for Japan, Denmark, France, the United Kingdom, and the U.S., 1880-1980.

Figure 4.1. Key

Argentina	AR	Norway	No
Australia	Aus	Pakistan	Pak
Austria	Au	Paraguay	Par
Bangladesh	Ba	Peru	Pe
Belgium (& Luxembourg)	Be	Philippines	Ph
Brazil	Br	Portugal	Po
Canada	Ca	South Africa	SA
Chile	Ch	Spain	Sp
Colombia	Co	Sri Lanka	Sr
Denmark	De	Surinum	Su
Egypt	Eg	Sweden	Swe
Finland	Fi	Switzerland	Swi
France	Fr	Syria	Sy
Germany, F.R.	Ge	Taiwan	Ta
Greece	Gr	Turkey	Tu
India	In	U.K.	UK
Israel	Is	Venezuela	Ve
Italy	It	Yugoslavia	Yu
Japan	Ja		
Libya	Li		
Mauritius	Ma		
Mexico	Me		
Netherlands	Ne		
New Zealand	NZ		

cultivation of the areas already used for agricultural production. Increases in food and fiber production will depend on continuous advances in agricultural technology. In countries with no adequately institutionalized agricultural research, farmers will be unable to respond to society's demands for food and fiber.

It is useful to distinguish between advances in mechanical technology, which generate increases in output per worker and advances in biological technology, which generate increases in output per hectare, (Hayami & Ruttan, 1985). Figure 4.1 depicts the long-term patterns of growth in output per hectare and output per worker for Japan, Denmark, France, the United Kingdom, and the U.S. between 1880 and 1980. The figure also identifies the current levels of both output per worker and output per hectare for a number of other developed and developing countries. The story that emerges from this illustration is that both Japan and the U.S focused their scientific and technical efforts toward eliminating the constraints on production imposed by scarce, or relatively expensive, factors of production. In Japan, labor was cheap and land was expensive. Japan, almost from the outset, directed its technical development toward increasing output per hectare. In the U.S., however, labor was expensive and land was cheap. The U.S directed its efforts primarily to expanding output per worker. During the last several decades, the Japanese and U.S. patterns have begun to converge.

The implication of the "induced technical change" model used to interpret the long-term technical changes illustrated in Figure 4.1 is that the direction of technical change is primarily endogenous to the economic system. This model was extended to interpret institutional change induced, to a substantial extent, by changes or differences relative to resource endowments and technical change (Hayami & Ruttan, 1985). In the induced institutional change model, advances in natural science knowledge reduce the cost of technical change, and advances in social science knowledge reduce the cost of institutional change. Cultural endowments may either reduce or increase the costs of technical and institutional innovation.[2]

The Agricultural Scientist as Hero and Villain

The productivity of modern agriculture is the result of a remarkable fusion of technology and science. In the West, it was built on ideological foundations that, from the early Middle Ages, have valued both the improvement of material well-being and the advancement of knowledge. This fusion did not evolve easily. The advances in tillage equipment and cropping practices in Western Europe during the Middle Ages and well into the 19th century evolved entirely from husbandry practice and mechanization insight. "Science was traditionally aristocratic speculative, intellectual in intent; technology was lower-class, empirical, action-oriented" (White, 1968; Asimov, 1979; Boulding, 1966; Hannay & McGinn, 1980). This cultural distinction has persisted in folklore about the priority of basic science over applied science long after the interdependence of science and technology had eliminated the functional and operational value of the distinction.

The power that the fusion of theoretical and empirical inquiry has given to the advancement of knowledge and technology since the middle of the 19th century has dramatically impacted the integrity of traditional institutions and non-natural environments. One can argue, not unrealistically, that agronomists, along with engineers and health scientists, have been the true revolutionaries of the 20th century.[3] Discovering heroic qualities in the pioneers who have carried banners for the agricultural revolution is not difficult. We can recall many examples:

- Liebig battled to establish the theory of the mineral nutrition of plants, and Mendel patiently distilled the elementary laws of genetics from the color of peas in his monastery garden.

- Harry Ferguson, the self-taught mechanic, applied basic physical principles to the integrated design of tractors and tractor equipment.
- Donald Jones had to escape from the orthodoxy of the corn breeding program at Illinois to the obscurity of Connecticut to have the freedom to explore the potential value of hybrid vigor.
- Vavilov, the great plant pathologist-geneticist-wheat breeder risked mind and body protecting the integrity of the Institute of Plant Breeding against the ideological opportunism of Lysenko.

But agricultural scientists have been *reluctant* revolutionaries. They wanted to revolutionize technology, but preferred to neglect the revolutionary impact of technology on society. They often have believed that revolutionizing agricultural technology can be done without changing rural institutions. Because they believed this, they frequently failed to recognize the linkage between the technical changes in which they took pride and the institutional changes which they either did not perceive or feared. They had internalized a simple utilitarian vision of moral responsibility—to make two blades of grass grow where one grew before (Thompson, 1988b). Then they have often reacted with shock and anger when confronted with charges of responsibility for institutional change—in labor relationships, tenure relationships, and commodity market behavior—that were induced by this technical change.

Responsibility for Research Results

How should the agricultural scientist or science administrator respond to these charges? Can they be dismissed as the mistaken or malicious rhetoric of romantics, populists, and ideologists?[4] How does one engage in useful dialogue about the role of science in society in an atmosphere that is politically and emotionally charged? A first step is to recognize that similar economic and social forces have generated both the drive for technical change—leading to advances in the productive capacity of plants, animals, machines, and human beings—and the drive for institutional change—designed to achieve more effective management of the direction of scientific and technical effort and capacity. The increased scarcity of natural resources, such as land, water, and energy continues to create demand for technologies that are capable of generating higher levels of output per worker, hectare, and kilocalorie. The increasing value society places on the health of workers and consumers, as well as on environmental amenities such as clean water, clean air, and clean streets, leads to a demand for effective social controls over the development and use of agricultural technology.

This enhanced sensitivity to the moral, aesthetic, and economic implications of technical change imposes expanded responsibility on both public and private decision processes. Greater responsibility is demanded for the way scientific results and technology are implemented. Should government respond to this demand by changing the institutions that induce the generation of new knowledge and new technology? Should government assume a stronger role in directing and limiting the adoption of new technology? Should it attempt to encourage greater aesthetic and moral sensitivity from scientists, engineers, agronomists, and science administrators? The difficulties that face governments in attempting to respond to these questions are formidable, but not insurmountable. Public demand for greater moral responsibility in the generation and use of new technology can be illustrated more concretely through specific examples. The 1970s controversy over the employment displacement effects of the tomato harvester serves as one useful illustration, and the case of research on tobacco improvement represents a second illustration. These examples depict, in a dramatic way, these pervasive moral principles.

Technical Change and Employment Displacement

The introduction of the machine harvesting of tomatoes has been accompanied by an especially vigorous debate. It has been considered the product of a uniquely effective collaboration between mechanical engineers and plant scientists (Rasmussen, 1968). At the same time, it has been vigorously attacked for displacing farm workers and small producers (Schmitz & Seckler, 1970; Friedland & Barton, 1975; Just et al., 1979).

In 1978, a suit on behalf of the California Agrarian Action Project and a group of farm workers was filed against the University of California Regents, charging that they had allowed agribusiness corporations and their own economic interests to influence the decision to spend public tax funds to develop agricultural machines. The relief sought by the plaintiffs included an order to compel the University to use the funds received from machinery patents to help farm workers displaced by those same machines. The farm workers' lawsuit has a long and intricate history which cannot be detailed here. Labor was joined by small farmers to bring claim against the University under a suit filed by California Rural Legal Assistance (CRLA). Ironically, the finding in fall of 1987 ruled on behalf of the small farmers who were harmed, but rejected the standing of farm laborers, who originally filed the suit.

The case attracted attention because of its ethical merits long before any judicial finding was issued, however. In December 1979, U.S. Secretary of Agriculture Bob Bergland announced that he intended to stop USDA funding for research that might put farm laborers out of work (Marshall, 1980). The dean of the University of California College of Agricultural and Environmental Sciences at Davis criticized Bergland for attempting to impose restrictions on the freedom of academic research. These two events symbolize the dilemmas faced by agricultural research administrators.

Clearly, the farm workers displaced by labor-saving machinery deserve a reasonable degree of protection from unemployment. This is a legitimate claim on the new income streams (productivity dividends) resulting from the adoption of the new technology. But who among the displaced workers deserves protection? Do the displaced workers who immediately found other employment have a legitimate claim on the new income stream? What about the workers who found other employment but at lower wages? And what about the tomato growers in Indiana and New Jersey who lost part of their market because of lower costs in California? These questions show that the *prima facie* validity of the farm workers' claims must be qualified by a series of complicating factors.

At the same time, any suggestion that University of California scientists must bear the responsibility for meeting the claims of the workers would impose a severe burden upon future research. In fact, other involved individuals might be expected to bear the responsibility to compensate those actually harmed by the introduction of the tomato harvester. For example, should the farmers who captured the initial gains from lower costs or the processors who increased their production because of the ability to expand their California operations be responsible? Or should it be the consumers who ultimately gained as competitive forces transferred the lower costs of production on to them?

The response is implicit in the questions. The gains of productivity are diffused broadly. And the costs should be borne broadly—in the form of general rather than specific protection. In a wealthy society such as the U.S., a worker should not have to prove specific displacement in order to be eligible for such protection. Whatever action may have caused displacement, responsibility for redress should rest with society as a whole. The first line of defense against the impact of displacement is an economy in which productivity is growing and employment is expanding. Society has little obligation to compensate workers who can readily find alternative employment.

The second line of defense is a program of severance payments and unemployment insurance that is effective for all workers—those who are forced to seek seasonal or casual employment as well as those in more favored industries.[5] A society that provides generalized protection will be in a stronger position to realize the gains from technical change, and to diffuse these gains broadly, than a society insisting on specific or categorical protection. Failure to develop institutions capable of protecting farm workers from the effects of seasonal unemployment and technological displacement has resulted in the transfer of an excessive burden of displacement costs on farm workers. This, in turn, has induced a legal and political response that, if effective, could slow technical change and limit the gains from increased productivity.

In a society in which employment opportunities are expanding rapidly and protection from unemployment is adequately institutionalized, neither the individual researcher nor the director of the research team involved in the development of a new tomato or lettuce harvester should be excessively burdened by the moral implications of trade-offs between economic and social costs and benefits of mechanization. Public policy has relieved them of that burden. But who should bear the burden of responsibility in a wealthy society that forces the burden on its poorest citizens? The moral question should be shifted from one addressed simply to agricultural research to one focused on social justice in the broadest possible sense.

Efficiency in the Production of a Health Hazard

Tobacco is a commodity that has been the subject of moral debate and political intervention since it first became a commercial export from colonial America. In the 1950s and 1960s, conclusive evidence associated cigarette smoking with lung cancer, coronary artery disease, chronic bronchitis, and emphysema. The source of the health hazards is nicotine and related alkaloids (U.S. Public Health Service, 1980).[6] What are the moral responsibilities of agricultural researchers and research administrators in relation to a crop that not only causes chemical dependency but also kills people or at least has a high probability of shortening the life of those who consume its products?

Under these circumstances, one would think that efforts to develop tobacco varieties with low nicotine content would have the support of both farmers and consumers. Yet in the early 1950s, a successful effort by Professor W. D. Valleau of the University of Kentucky to develop low-nicotine varieties of tobacco was bitterly attacked by Kentucky farmers because of potential competition with burley tobacco (Hardin, 1955). In retrospect, we have little difficulty in sympathizing with the objectives of Professor Valleau's research. Even a marginal contribution to the reduction of chemical dependency and the health hazards of cigarette smoking would seem to be desirable today. But what about the issue that underlies this judgment?

Should public funds be used to conduct research that reduces the costs and improves the productivity of a product that causes chemical dependency or shortens life expectancy? What are the moral responsibilities of the directors of the agricultural experiment stations in the states that support tobacco research? And what about scientists who devote their lives to understanding the physiology or the nutrition of the tobacco plant? Is the farmer who grows the tobacco absolved from responsibility because of the market demand for tobacco? Are members of the legislature and the experiment station directors absolved because tobacco has been one of the more profitable crops available to small farmers in the depressed areas of Kentucky or North Carolina? Are the scientists relieved of responsibility by an appeal for the freedom to do research? What are the moral implications for the tobacco breeder, whether employed by a private firm or a public research institution, for responding to market criteria when the market is most effectively enhanced by producing a crop that causes chemical dependency?

What inferences can be drawn about the moral responsibility of a society in which the government spends billions of dollars on medical and health care because of smoking, on research on tobacco-related diseases, and on campaigns to discourage smoking? This same society supports research to improve efficiency in tobacco production and legislates programs to support the incomes of tobacco producers. As in the case of the tomato harvester, institutional changes can relieve research administrators and scientists of the moral dilemma posed by tobacco research. If public consensus made the sale of tobacco products illegal in the U.S., the directors of the Kentucky and North Carolina state agricultural experiment stations would not likely allocate any more resources to tobacco improvement than they now allocate to marijuana research. So far, no sufficient convergence of opinion has formed to take the steps necessary to limit the content of dependency-forming or carcinogenic substances in cigarettes. An attempt to move toward complete prohibition would require a careful balance between the desirable effects on individual health and the undesirable effects of enforcing prohibition.

Toward Some Guidelines to Moral Responsibility

The tomato and tobacco research cases pose extremely difficult moral problems for agricultural researchers and research managers. The centuries-long struggle in western society to free scientific inquiry from the constraints of the church make it unlikely that the answers to issues of moral responsibility for new knowledge and new technology will be sought from traditional religious sources.

Where, then, can the scientist or science administrator look for guidance on issues of moral responsibility? One possibility is a philosophy of inquiry approach that recognizes the objective status of both positive and normative knowledge.[7] The philosophy of scientific inquiry to which most scientists subscribe, either explicitly or implicitly, imposes only two criteria as a test of *objective knowledge: correspondence* and *coherence.* The test for *correspondence* requires that knowledge be continually tested against experience and observation. The test for *coherence* requires that scientific explanation meet the test of *logic;* it must be explainable in terms of our general knowledge of scientific principles.

This view of scientific method, known as logical positivism, has been of great significance in quantifying scientific knowledge. Biometricians, econometricians, and others can clearly distinguish between the logical structure of their concepts, which can be tested for coherence, and the empirical content of their statements, which can be tested for correspondence. Logical positivists of the 1930s maintained that statements which could not be tested for empirical content were meaningless. Although this principle has been essentially abandoned by philosophers, primarily on logical and scientific grounds, it continues to influence scientific thinking even today. On this foundation, logical positivists concluded that there was no empirical, objective, or true knowledge of the normative.

The recent social criticism of science can be interpreted as social pressure for paradigm change (Kuhn, 1962). Some philosophers of science draw an analogy between the tentative "dialectical" nature of positive scientific knowledge, which must be continually tested for correspondence with empirical observations, and the tentative nature of normative knowledge. They argue that normative experience, such as the goodness of a healthy body or the badness of injustice, implies that normative knowledge can also, in principle, be tested against the criteria of coherence and correspondence. Like positive knowledge, normative knowledge is always tentative and must be continually tested and retested for correspondence with social behavior.

Acceptance of comparable objectivity of positive and normative knowledge does not, however, lead directly to prescriptions about right or wrong behavior. Doing what is bad is not always wrong—if it is the least bad thing that can be done under the circumstances. Nor is it always right to do what is good if something even better can be accomplished with the same, or with less effort or resources. Thus, the knowledge that cigarette smoking has bad effects does not automatically imply that cigarette smoking should be prohibited. A decision to prohibit cigarette manufacture and trade would involve the good effects of less smoking on chemical dependency, health, and longevity against the bad effects, such as corruption of the legal system and loss of personal freedom.

Any decision that transforms knowledge about what is good or bad into a prescription about what should be done implies the use of both normative and positive knowledge. But agricultural scientists have always applied such a rule to their research. They have simply internalized fairly uncontroversial normative assumptions concerning the goodness of producing more food ("three blades of grass where one grew before") and the drudgery of the most onerous forms of farm production. Such decisions should involve both positive and normative knowledge about consequences, not only for agricultural production, but also for the health of humans, the incomes of hired laborers and farm operators; the cost of food to consumers; and the economic and political status of scientists, bureaucrats, and politicians. Only modest progress has been made in creating a set of tested normative knowledge that can serve as a basis for workable prescriptions. Yet one can perceive, in the public discussion of the tomato harvester and tobacco research cases, three principles that appear to have fairly broad applications in interpreting a wide range of individual and group behavior in issues of moral responsibility.

The first is that a risk or loss that occurs incrementally is associated with less personal concern and creates weaker public response than a risk or loss that occurs in more discrete or lumpy units. For example, most smokers seem willing to accept the statistical evidence that cigarette smoking reduces average life expectancy. But almost no one believes that smoking one more cigarette or waiting one more week to stop smoking will have an effect on his or her own life expectancy. In contrast, the loss of a job as a result of the mechanization of a harvest operation or the displacement of domestic employment by imports is a discrete and often very painful event. Ethically, the principle is that a catastrophic loss (e.g., one that happens all at once) is worse than an otherwise comparable loss that occurs incrementally. A catastrophic loss often generates a substantial political response, even when the number of individuals affected is relatively small.

A second principle in determining response to moral responsibility issues is asymmetry between gains and losses. The utility gained by an incremental increase in income or by the control over an environmental amenity is less than the utility lost from an equal incremental loss in income or environmental amenity. For example, a loss of $10,000 would create severe hardships for most middle class families. A sudden windfall of $10,000, however, might only mean that a new car is purchased a little earlier than planned. This asymmetry leads to stronger political action to prevent income or quality deterioration than action to achieve income on quality gains (Hardin, 1982).

A third principle is that people are willing to tolerate a greater level of voluntary risk than involuntary risk (Rescher, 1983). Government has a clear responsibility to protect citizens against damage or loss imposed on them by either the purposeful or unintended actions of others. In contrast, most believe that government had only limited responsibility for protecting citizens against the damage they do to themselves. Smokers who are willing to acknowledge the effect of their habit on life expectancy may also insist that smoking is a matter of personal choice and may be willing to defend tobacco research that will provide less expensive or better cigarettes. This principle is derived from the importance of consent in establishing the social contract. Strong versions of the contract principle maintain that government has no right or responsibility to interfere in the actions of consenting adults.

These three principles appear to correspond to the majority of personal and group behavior. The continuous testing and evaluation of the values implicit in individual and group behavior represent an important element in our attempt to derive the normative knowledge needed to develop workable prescriptions in the field of science policy.

Technology and Reform

A necessary step in any effective response to public concern about the social impact of technical change is for the research community to agree that society has a right to hold scientists, engineers, and agronomists responsible for the consequences of the technical and institutional changes caused by their research.[8] When credit is claimed for the productivity growth generated by advancements in agricultural technology, responsibility cannot be evaded for the effects on the distribution of income between the suppliers of labor, land, capital, or industrial inputs. Nor can responsibility be evaded for impacts, such as pest control chemicals on environmental amenities or on the health of workers and consumers.

Once the question of society's right to make scientists, engineers, agronomists, and economists responsible for the effects of the knowledge and technology they provide is accepted, the more tractable question can be examined: What responsibility will a wise society impose on its research community? In this case, *it is in society's interest to let the burdens of responsibility rest lightly on the shoulders of individual researchers and research managers.* If society insists on being ensured that advances in agricultural technology produce minimum risk (that agricultural scientists abandon their revolutionary role), society must accept the risk of losing new income streams generated by technical change.

Critics and political leaders should, for example, exercise extreme care in suggesting that research managers and scientists commit themselves to the realization of scientific or technical objectives that are unrealistic in terms of the state of scientific and technical knowledge. In the 1950s, for example, expecting that utilization and marketing research could make a significant contribution to the solution of agricultural surplus problems in the U.S. was unrealistic. The allocation of excessive research resources to these areas led to both a waste of research resources and to erosion in the credibility of the research enterprise.

Asking agricultural research managers and scientists to adopt objectives that do not correspond to the economic or political marketplace is equally wasteful. It is unrealistic, for example, to insist that the California Agricultural Experiment Station direct its mechanization or biological research to the needs of the 160-acre farm—unless the State of California or the federal government is prepared to support the structural policies necessary to reverse the trends toward large-scale agriculture. A research system cannot be asked to produce knowledge and technology that will not be used without eroding the intellectual integrity, and ultimately the scientific capacity, of the research system.

Some may argue that policy-makers should insist that research managers direct social science research to "discover" society's true objectives (a social welfare function) prior to the time that it is "revealed" in the political or economic marketplace. This implies that research managers should have analytical expertise on their staff, not only to assess the benefits and burdens of technical changes anticipated from a research program, but also the capacity to develop a set of normative weights (shadow prices) that reflect the "true" value society places on the welfare of individuals or groups that may potentially benefit or lose from the research results. The information could then be used to make research resource allocation decisions. This viewpoint suggests that research directors should allocate resources on the basis of a social welfare function prior to the time research results are revealed by either the economic or political system.

How do research managers respond to the social concern about the impact of technology on society? Research directors should have access to the analytical capacity to gauge the potential incidence of benefits and burdens. This enables them to maintain effective dialogue with the political system about research budgets and priorities. The research director who does not have access to or fails to use such capacity stands naked before critics and supporters. Research leading to a better understanding of the discrepancies or the disequilibrium in the economic, political, and social weighting system is essential. *But the objective of such research should not be to provide research directors with a decision rule system for internal research resource allocation.* The objective, instead, should be to contribute to a political dialogue that will result in institutional changes leading to convergence of the economic and political weighting systems. As these weighting systems meet, research directors will not be forced to choose among alternative responses to an arbitrary or inconsistent set of economic, political, and social weights.

Research managers do have a clear responsibility to inform society about the impact of pricing systems and tax structure on (1) the choice of mechanical, chemical, and biological technologies by farmers; (2) the incidence of technical change related to distribution of income among laborers, landowners, and consumers; (3) the structure of farming and rural communities; and (4) the health and safety of producers and consumers. They also have a responsibility become involved in intellectual and political discussions that allow society to achieve more effective convergence between "market" and "shadow" prices and between "individual" and "revealed" consumer preferences.

If market and shadow prices for inputs and products can be made to converge, research directors will have clear signals about how to allocate research resources. When market and efficiency prices diverge, influencing research planners to allocate resources consistent with shadow prices is almost impossible. If political processes can attain greater consistency between "revealed preferences" and normative values, scientists and research managers might not be confronted by situations such as the tobacco case: Although cigarette smoking has been branded as dangerous to health, new public resources are continually appropriated for tobacco research.

A major qualification should be mentioned here. Few research directors make adequate use of the knowledge available within their own institutions to learn about either the positive knowledge on what is scientifically and technically feasible, or the normative knowledge about potential value of new knowledge and technology. And, as a whole, agricultural institutions continue to evade their responsibility to engage in public dialogue about the costs, as well as the benefits, of technical change in agriculture. This issue should be a major focus of public affairs extension.[9]

Agricultural Research and the Future

What should society expect from agricultural science in the future? And what does agricultural science have a right to expect from society if it continues to meet society's expectations?

First, society should insist that agricultural science maintain its commitment to expand the productive capacity of the resources used in agricultural production. These include the original endowments of nature (soil, water, and sunlight) the agents humans have domesticated or adapted (plants, animals, and organic and mineral sources of energy) the agents invented (machines and chemicals) and the individuals engaged in agricultural production. By the end of this century, the capacity to maintain this commitment should be established in every part of the world. The world continues to depend on the productive capacity of North American agriculture. This dependence poses danger both to the developing world and to North America. Effective agricultural science communities and institutions capable of producing the knowledge and technology to reverse the trends of the last several decades must be established. Agricultural science in North America must remain strong enough and sufficiently cosmopolitan to contribute to and learn from the emerging global agricultural science community.

Second, society should insist that agricultural science embrace a broader agenda that includes a concern for (1) the effects of agricultural technology on the health and safety of agricultural producers; (2) the nutrition and health of consumers; (3) the impact of agricultural practices on the aesthetic qualities of both natural and manmade environments; and (4) the quality of life in rural communities. Research planners should consider the implications of current technical choices for the options that will be available to society in the future. These concerns are not new for agricultural science. But they often have been viewed as peripheral or diversionary to its primary task. The agencies responsible for the financial support of agricultural research too often have failed to look beyond the farm gate. To ensure a successful future for agricultural science, these concerns must be fully embraced. The capacity to work on these problems outside of the traditional agricultural science establishment also must be maintained so an effective dialogue can be achieved both within the research community and in the realm of public policy.

Now what should the agricultural science community expect from society? First, agricultural science should expect that each citizen, in general, and political leaders, in particular, will gradually acquire a more sophisticated perception of the contribution of agricultural technology to the balance between humanity and the natural world. The position that agricultural science is engaged in a continuous assault on nature is as wrong as the attitude that agriculture is a bastion of moral purity. Society must understand that agricultural science can succeed in expanding productive capacity only as it reveals and cooperates with the laws of nature. In the Western World, we are the inheritors of a tradition that considers material concern as a defect in human nature.

This inheritance leads to a romantic perspective of humankind's relationship to the natural world. It also leads to a viewpoint that technology alienates humans from both the natural world and the natural community. Scientists, engineers, and agronomists have a right to expect the philosophers of society to achieve more sophisticated insight into our relationship to technology and nature. We must recognize that the invention, adaptation, and use of knowledge to achieve material objectives does not "reduce" experience, but rather expands it.

Second, it is time for the general science community to follow the lead of the agricultural science community in embracing the fusion of science and technology, rather than continuing to hide behind the indefensible intellectual and class barriers. The ideal of "pure" science has been advanced to protect its privilege and its ego from contamination by engineering, agronomy, and medicine. This change will become increasingly important in the future as the close of the fossil fuel frontier joins with the close of the land frontier to drive technical change along a path which implies a much larger role for biological and information technology.

The 1970s was a period of declining productivity growth in the U.S. as well as several other advanced economies. These dangerous trends were more apparent in the industrial than in the agricultural sector. Rates of return on agricultural research have remained high. The evidence suggests that institutional linkages providing effective articulation between science, technology, and agriculture have continued to be productive sources of economic growth in both developed and developing countries. Much can be learned from this experience if we are not blinded by outmoded status symbols or cultural constraints.

Notes

1. Some interpret the mechanization of land preparation or harvesting as a source of poverty in rural areas rather than as a response to rising wage rates. Grain milling by wind and water power was considered progress in 12th century Europe. But today' critics view the substitution of rice mills for hand pounding as eliminating job opportunities in 20th century Java. Some regard the use of fertilizers to increase food production as poisoning the soil rather than removing the pressure of agricultural production on marginal lands and fragile environments. The new income streams that flow from more productive farms are considered destructive to the integrity of rural communities rather than as ways to enable rural citizens to participate in a society in which the gap between rural and urban income, lifestyles, and culture has narrowed.

2. A large collection of social science literature views cultural endowments as constraints on technical and institutional change. The literature also contains

numerous cases supporting the perspective that when the design of technical or institutional innovations are compatible with cultural endowments, traditional institutions can be important resources for change (Ruttan, 1988).

3. The term "agronomy" here refers to the whole body of agriculturally related science and technology, and "agronomists" refers to the community of production-oriented agricultural scientists. In the U.S., the term agronomy has the more narrow connotation of field crop production and management.

4. For an example of romantic criticisms, see Berry, 1977. For a populist perspective, see Hightower, 1973. For an ideological perspective, see Lappe and Collins, 1977. Also see Valliantos, 1977, and George, 1977. For reviews of this literature, see Eberstad, 1980, and Hardin, 1979.

5. Hayek has argued that in a market economy, the concept of social or distributive justice has no meaning. "Justice has meaning only as a rule of human conduct and no conceivable rules for the conduct of individuals supplying each other with goods and services in a market economy would produce a distribution which could be meaningfully described as just or unjust" (Hayek, 1978). Hayek contends, in effect, that justice is a function of the rules or processes given individual and group behavior and not of the outcome generated by the rules. My interpretation of the social contract between labor, capital, and consumers in the U.S. is that when markets function in a workable manner, sorting out the distributional effects of technical and institutional changes associated with economic growth are left to the market. But if the market is judged ineffective or inequitable in partitioning the benefits and costs of economic growth, alternative institutions, such as unemployment compensation and social security, will be designed to achieve equity objectives.

6. Prohibition of tobacco use would undoubtedly not be any more effective than attempts made in the 1920s in the U.S. and Finland to prohibit the consumption of alcoholic beverages. Of interest, however, are the severe restrictions imposed in Malaysia in 1980 on advertising alcoholic beverages and cigarettes. Advertising restrictions on tobacco have been less severe than those on alcoholic beverages, primarily because tobacco is produced by large numbers of small farmers.

7. For a discussion of these issues in the context of energy policy, see Johnson and Brown, 1980.

8. Mohr (1979) notes that "... freedom of inquiry ... does not necessarily imply freedom in the choice of any particular goal; it implies, however, that the results of scientific inquiry may not be influenced by any factor extrinsic to science."

9. An important exception is the excellent effort by Cornell University to assess the import of the bovine growth hormone (bsT) and to extend that knowledge to the New York State dairy industry.

5

An Agroecological Analysis of the Environmental Degradation Resulting from the Structure of Agriculture

Miguel A. Altieri

Introduction

Agricultural systems are the result of the coevolution of ecological and social processes because agriculture requires converting natural resources, including plants and animals, into useful products through the application of human knowledge and material resources. These resources are made more productive through the use of capital (infrastructure, inputs, equipment, credit, etc.) (Norgaard, 1984). Ecologically speaking, agriculture is the artificialization of nature. Understanding the environmental consequences of such transformation to meet the evolving demands of human populations is the primary concern of environmentalists (Altieri, 1987). Agricultural management implies a set of complex relationships between land, capital, labor, policies, and institutions. A goal of social scientists is to realize such complexity and appreciate the difficulties of achieving a "good fit" among the various factors, processes, and institutions (Uphoff, 1986).

Understanding how economies of scale, technological change, and agricultural policies divert agricultural production away from basic needs and resource conservation toward what is profitable in domestic or international markets (and not necessarily environmentally sustainable) is perhaps the most important challenge facing social and environmental sciences today. The task is difficult because the relationship between agricultural development and environmental preservation is compounded by several factors.

- The large number of farms involved, each with their own set of human, natural, and material resources; socio-cultural backgrounds; decision-making processes; constraints; etc.
- The large variations in physical, ecological, and socio-economic environments within which farms operate.
- The variety of institutions, facilities, and activities involved in agriculture which provide technical assistance, extension, credit, marketing, etc.
- The difficulties in establishing the boundaries of agricultural development, both in terms of its ecological and socio-economic impacts (i.e., at the individual, household, community, or regional level).
- The tendency of large-scale agriculture to externalize production costs, most of which are absorbed by society at large, rather than by the individual farmers. Thus, environmental costs become hidden costs, rarely accounted for by economic cost/benefit analysis.

The main argument of this chapter is that given such complexity, the analysis of the environmental crisis associated with agriculture cannot be viewed by disentangling environmental from social and economic issues. On the contrary, the agroecological paradigm maintains that both must be examined holistically because of the following reasons:

- Environmental degradation associated with agriculture is not only an ecological process, but also a social and political-economic process (Blaikie, 1985).
- The determinants of environmental quality are consequences of agricultural structure and policy. Therefore, the environmental problems of agriculture are inherent to the major structural features of the prevailing agroeconomic system (Buttel, 1980a).

Structure of Agriculture

The structure and organization of agriculture in any region or country evolve from of a set of socially defined historical forces, particularly markets, competition, and political power which determine the social control over resources and their productive use (de Janvry, 1982). A clear example is the historical dominance of California agriculture by large landowners who have developed a unique ability to influence federal and state policies regarding water, labor, markets, and mechanization, to maintain economic advantages (Liebman, 1983). At the world level, three fundamental processes have shaped the current structure of agriculture (de Janvry, 1982).

Commodification of Agriculture

Agriculture and its means of production have become a commodity. As a result, land and other natural resources, agricultural products, labor, and technology all become commodities that can be bought and sold in markets where they acquire a price. Such commodification and monetarization of agriculture has been fueled by the privatization of land and formation of labor, capital, and technology markets, which lead to production specialization and other capitalist social relationships in agriculture. The result is that economical rather than ecological considerations have increasingly defined what is produced, how it is produced, and for whom it is produced.

Agribusiness Dominance Over Farming

Farming has integrated into and submitted to agribusiness, thus allowing agribusiness to gain control over farming through monopolistic supply of inputs, vertical integration, and contract farming (Vogeler, 1981). Despite this control, capital has not taken over the farms themselves because purchase of land ties up large amounts of capital, and the turnover rate of capital is limited by the annual cycle of growth. More importantly, however the labor process is difficult to supervise and control. Thus, production and the associated risks are the responsibility of independent producers, while the labor process is determined from above and outside (Levins & Lewontin, 1985).

Agribusiness dominance over farming has also introduced specific biases about production, farm size, and the social class structure in agriculture, primarily through direct involvement of the private sector in agricultural research, as well as through its influence over public research (Friedland et al., 1981). This research, clearly designed to maximize the use of purchased inputs, has become a dynamic force of new technologies tailored for large farms and focused on commodities with high effective demand. A clear example is the mechanization of tomato harvesting, which displaced a large number of small farms and redefined the size of small farms to a minimum of 200 acres (de Janvry et al., 1980).

Internationalization of Capital

Capital has internationalized in agriculture, as manifested in foreign investments in plantations or contracts between agribusiness and foreign producers, as well as establishment of new markets for agricultural exports. The result is that agricultural systems throughout the world have become so integrated that the logic of local agriculture is increasingly dictated by world market and international political forces (Schuh, 1976). The internationalization of capital is diverting agricultural production away from basic needs toward what is profitable in the market. The technology used also caters to existing profit-making opportunities, increasingly emphasizing labor and land-saving chemical and mechanized technologies.

Environmental Crisis in U.S. Agriculture

After World War II, the process of farm consolidation and dominance of agriculture by agribusiness and large owners in the U.S. accelerated dramatically. These trends have led to the current structure of agriculture, which is characterized by (1) outmigration; (2) concentration of land, resources, and production; (3) increase of capitalist farming; and (4) biased policies that encourage farm size expansion, specialized production, and high levels of mechanization and high-input use (de Janvry & Vandeman, 1982).

The structure of California agriculture has been shaped by an unprecedented increase in irrigated acreage of 2.7 million acres (about 90 percent of the harvested cropland), and intensified production, resulting in enhanced output per acre for most products. The need for labor due to mechanization and use of chemicals has noticeably decreased (about a 25 percent decline in the need for seasonal laborers), and the number of farms with a concomitant rise in farm size also has declined. Most of these changes have been fueled by the increased ability of large-scale producers to gain political power, thus affecting state and federal policies on water, labor, mechanization, and public research issues. One illustration is the increase in production of cotton, rice, and sugarbeet production made possible by influencing the federal government's commodity programs to provide good markets. Other examples are the University of California's research on mechanization, and the establishment of regulations for unrestricted access to sources of water through the State Water Project (Liebman, 1983).

Clearly, profit motivations, rather than environmental concerns, have decidedly influenced the forces that shaped the structure of agricultural production in the U.S. (Cochrane, 1979). This fact alone constitutes a major barrier to more environmentally sound management of agricultural resources. Thus, the very nature of agricultural structure and associated policies lead to the environmental crisis affecting agriculture (Buttel,

1980b). Each structural feature affects environmental quality in several ways. For example, farm size encourages specialized production and mechanization, which, in turn, leads to agroecosystem degradation associated with monoculture, a production system with increased reliance on chemical inputs and abandonment of conservation practices such as rotations. Similarly, substitution of capital for labor increases energy intensity, encouraging inputs to the point of energy inefficiency, and facilitates specialization, which increases the risk of system failure because of genetic and horticultural uniformity (Adams et al., 1971).

Environmental problems in U.S. agriculture (i.e., land degradation; salinization; pesticide pollution of soil, water, and food chains; depletion of ground water; genetic homogeneity and associated vulnerability) are the result of a "technological saturation" associated with the intensification of production using chemical inputs. Thus, the environmental problems of agriculture should be perceived as symptoms of an "ecological disease," with roots that are inherent to the structural features of an energy-intensive, large-scale, specialized agricultural industry, rather than events that accidentally occurred.

Agriculture and Environmental Quality in the Third World

Because of tendencies in the global economy and pressure from international lenders and foreign governments, most developing countries are turning to agricultural exports as an economic strategy to save the debt and foster development, which has been extremely uneven at best (World Bank, 1986). The results have been bankruptcy of the peasantry, a loss of food self-sufficiency, and increased dependency on technological inputs, many of them restricted or banned in industrial countries because of negative environmental effects (Barkin & Suarez, 1982).

Most Third World countries are experiencing a deepening crisis, with populations rising, land holdings shrinking, environments degrading, and per capita food production remaining static or declining. This crisis affects mostly the resource-poor farmers and those dependent on them—some 1.5 billion people. Although a connection exists between rural poverty and environmental degradation, the poor are only agents of other economic and structural forces and policies that push them to invade and farm fragile environments. As a result, they abandon the traditional systems of agriculture which are rich in elements of resource conservation. Nevertheless, the environmental impact of the masses of rural poor is low compared to the damaging effects of the economic activities of the large landowners and corporations. Their overuse or careless use of mechanical technology and agrochemicals creates major ecological threats (de Janvry & García, 1989).

In the Third World, environmental degradation associated with agriculture is clearly a symptom of underdevelopment and reinforces that condition (Blaikie, 1985). The major environmental problems are soil erosion, pesticide pollution, deforestation, and genetic erosion—all processes linked to overexploitation of natural resources. This overuse is because of poverty or massive transformation and perturbation of the environment in areas subjected to export agriculture, or recent colonization. Solutions for major environmental problems in these countries first requires a solution to the problem of social inequality and rural poverty. For soil, forest, and genetic conservation practices to be effectively implemented in these countries, fundamental social change first must take place (Blaikie & Brookfield, 1987).

Reverting Environmental Degradation in Agriculture

In the last two decades, several strategies have been proposed to promote change in agriculture. They are aimed at reversing the rise of environmental and social costs of agricultural modernization (Buttel, 1980a). The focus of these strategies has been the development of "appropriate technologies" of low environmental input, as if these technologies in themselves have the capability of initiating beneficial social changes. This technological determinism has, to a significant extent, prevented environmentalists from understanding the structural roots of environmental degradation linked to capitalist agriculture.

The relationship between environmental quality and prospects for increased food production is perhaps the greatest preoccupation of agricultural scientists today. Sustainable agriculture is being proposed as a strategy to translate the general principles of ecological resource management into production systems capable of balancing economic growth and natural resource conservation (Douglas, 1984). Although the strategy is novel in terms of bringing a long-term dimension and a systems approach together in the study of agriculture, it also suffers from technological determinism. Sustainable agriculture emphasizes low-input technologies as if their diffusion could cause a socioeconomic transformation of agriculture in and of themselves.

Sustainable agriculture, as depicted by mainstream institutions, has a relatively benign view of capitalist agriculture. It assumes the persistence of private property in agriculture and seeks to perfect it and/or make it more equitable. This narrow acceptance of the present structure of agriculture as a given condition restricts the real possibility of implementing alternatives that challenge such a structure (Levins, 1973). Thus, the possibility of a diversified agriculture is inhibited by the present trends in farm size and mechanization. Implementation of such mixed agriculture would only be possible as part of a broader program that includes, among other strategies, land reform and redesign of farm machinery. Otherwise, merely introducing alternative agricultural designs will do little to change the underlying forces that led to monoculture production, farm size expansion, and mechanization in the first place (Burbach & Flinn, 1980).

Most proponents of social change in agriculture have focused their efforts on (1) curbing or eliminating corporate ownership of farms; (2) protecting and encouraging small, family-sized farms; (3) developing appropriate and low-input technologies; and (4) promoting regionalism in food production and agricultural self-sufficiency (Merrill, 1976; Vogeler, 1981). The underlying conviction of these proposals is that a sustainable agriculture is incompatible with large-scale, corporate agriculture. Unfortunately, these proposals offer few insights on how to change production relationships in agriculture, other than a vague desire to return agriculture to an idealized state of a system of modest-scale farms, using low-input technologies and regional markets (Buttel, 1980a).

The expectation that a set of policy changes could be implemented for a renaissance of the family farm in the U.S. is unrealistic because it negates the existence of scale in agriculture and ignores the political power of agribusiness corporations and large-scale farmers. A more radical transformation of agriculture must emerge from political struggles that resolve (1) the conflicting economic demands of two classes: capitalist farmers and farm workers; and (2) the dilemmas created by inserting agriculture into the dynamics of the world economy and the international division of labor (de Janvry, 1982). In other words, change toward a more socially just, economically viable, and environmentally sound agriculture should be the result of social conflicts and social movements in the rural sector, rather than the diffusion of technological innovations, which lack power to bring about more equitable social arrangements.

Implementing Social Change and Technological Innovation

The structuralist perspective perceives environmental degradation as the consequence of the careless use of agricultural technologies and practices within the context of the current agricultural structure and policies. The present system sponsors specialization, farm size, capital and resource intensification, mechanization, and absentee ownership (Buttel & Gertler, 1982). Moreover, the structuralist perspective poses that the logic of local agriculture can no longer be understood in isolation from world/regional market forces and international/national political forces (de Janvry, 1982). In other words, the participation of each region and/or country in the world market of agricultural commodities has a major influence on the specific agricultural policies and structure of that region or country. Although realistic, the problem with this perspective is that with the complexities of implementing change at the national or international level, an element of hopelessness is introduced by suggesting that no change in production relationships can be expected from local action.

Recent experiences in the Third World suggest, however, that development in the right direction is possible, especially when implemented at the local level (Uphoff, 1986). A number of grassroots rural development programs in Latin America established in small peasant communities are broadening the base of participation in the production process under an agroecological paradigm, and are creating equitable arrangements for distributing the benefits from sustainable production increases (Altieri & Anderson, 1986). The focus has been to utilize "social energy" to promote a change of attitude that allows individuals to be responsible for their own development and therefore succeed collectively (Hirschman, 1984). Through local participation, local technologies are developed based on local knowledge and resources, and used as an instrument of liberation and social organization. This allows peasants to meet their basic needs without becoming dependent on external assistance. The objective is to build upon the best of the local initiatives to ensure and support self-reliance and local adaptation (Richards, 1985).

Evidence indicates that local successful experiences can occur regardless of whether social change occurs at the national level. This does not mean, however, that the sum of local-level changes will result in change at the regional\national level even though small farmers are capable of making changes that benefit them, as well as society as a whole. The concept of social energy contends that the resulting rise in consciousness that emerges from a successful local experience can fuel attitude changes on a much broader scale (Hirschman, 1984).

Rethinking an Agroecological Research Agenda

In the U.S., farmers constitute a unique sector of producers who own some of the means of production, but whose conditions of production, as argued in this chapter, are completely controlled by the monopolistic suppliers of inputs and purchasers of outputs. Based on the structural conditions, the agricultural research establishment, by serving the proximate demands of farmers, is actually an establishment that serves the needs of capital (Levins & Lewontin, 1985). This is why the focus of technological research has evolved from a problem, social-welfare-oriented focus toward a client, market-oriented one.

This issue is of key importance today with the advent of biotechnology, which is slowly exposing universities and research institutions to new and increased pressures to serve private interests at the expense of public interests.

The danger of this issue is that publicly funded agricultural research, like corporate research, may be targeted toward the development of profitable products to sell to farmers, rather than toward research to help farmers cut inputs and costs that affect the social goals of rural people, (Hassebrook & Hegyes, 1988). What may be lost in this process is research which addresses environmental problems by helping farmers develop and adopt resource-conserving agricultural practices. The problem is that most of these practices (i.e., biological control, organic soil management, soil and water conservation, etc.) offer minimal profit potential to agribusiness corporations, a constraint that cannot be ignored given the existing economic forces that drive agriculture.

As long as economic and political forces shape the research agendas of public institutions, research efforts will continue to be limited to "putting out fires." In other words, research is reduced to incremental modifications of existing technologies rather than to the generation new agroecological knowledge needed to guide the design of agroecosystems that emphasize long-term sustainability instead of short-term productivity (Buttel & Gertler, 1982).

Although most agricultural research has resulted in the development of technologies, many of which exacerbate problems in agriculture, a portion of the research efforts have been directed at monitoring the reduction of cropland productivity due to environmental degradation triggered by these technologies. In effect, the system creates its own "ecological watchdog," charged with assessing the environmental consequences of technology-induced environmental degradation. Most of the time, this uses resources and expertise otherwise needed to develop new alternative management systems that could offset such degradation.

The few new federal and state initiatives that foster research in sustainable agriculture are calling for proposals that focus on alternatives to energy- and chemical-intensive farming methods, but that increase real productivity (the ratio of output per unit of input) to maintain the continued financial stability in agriculture. "Low-input methods" of farm production are receiving increasing attention as a major part of the solution. This type of research, although important in reducing the use of farm chemicals and controlling soil erosion in specific farm situations, is expected to have a qualitatively insignificant effect on the structure of the agricultural sector (i.e., export orientation, taxation, farm size, land ownership, etc.), or on dominant farm practices (i.e., large-scale monocultures, mechanized operations, etc.)(Buttel, 1980a; 1980b). The fundamental dynamics of these structures lie elsewhere: in the world market, in inequality of land ownership, and in agricultural policies—not in the research sector. Agricultural research can contribute to change, but agroecological research agendas must include socioeconomic, political, and institutional innovations, as well as technological innovations (Uphoff, 1986).

This new agroecological research agenda must emphasize long-term solutions that address causes rather that symptoms. The challenge is to examine the relationships between agriculture, environment, and society in their broadest context by considering the natural and social processes to be intertwined. To develop an agriculture that is more ecologically and socially rational, research must accomplish the following (Levins & Lewontin, 1985; Altieri, 1987).

- Integrate ecological land use into agricultural planning so the boundaries of the agroecosystem are expanded to include naturalized ecosystems and the services they provide.
- Favor agricultural planning that pursues multiple goals such as food production, environmental protection, public health, resource conservation, etc.
- Focus on the development of gentle, thought-intensive technologies aimed at finding ways to reduce inputs instead of creating new inputs.
- Promote the development of alternative trajectories for crop management which can be sustained over the long-term based on economic, environmental, and social standpoints.
- Tailor technologies to specific agroecological and socioeconomic situations to maximize efficiency and match development to the needs and constraints of local farmers.
- Combine indigenous knowledge systems with the knowledge developed at research centers. This approach requires accepting local agricultural skills as the main resources for agricultural development.
- Emphasize that the development of environmentally sound technologies must occur within the context of a much broader rural development program and a particular direction of social change.

Finally, the study of the environmental dimensions of agriculture must be guided by a broader understanding of the relationship between natural resources, political institutions, economic forces, social classes, and the scientific community. This holistic understanding is crucial to realizing the following:

- Environmental degradation evolves from fundamental structures in society (Blaikie, 1985).
- A critique of these fundamental structures from an environmental point of view alone is not sufficient.
- All approaches to agricultural sustainability are ideological and underpinned by assumptions, conflicts of interest, etc.
- More often than not, the call for conservation occurs when degradation seriously affects the interests of the powerful classes.
- An ethic of social change must be adopted prior to any consideration of sustainable agricultural research and development (Blaikie, 1985).

6

Making the Land Ethic Operational: Toward an Integrated Theory of Environmental Management

Bryan G. Norton

Introduction

Aldo Leopold led two lives. He was, in the best tradition of Gifford Pinchot, a forester and a coldly analytic scientific resource manager devoted to the maximization of resource productivity. But Leopold was also a romantic. He joined the forest service because of a love for the outdoors, a love he never lost or fully subjugated to the economic "ciphers" that so constrain public conservation work.

During the last decade of his life, Leopold-the-romantic fashioned a little book of essays. He chose, from the best of his stacks of field journals and voluminous publications, a few short essays, supplemented these with new pieces, polished them, and strung and restrung them like little pearls. The manuscript, representing the essence of his long career as a forester and professor of game management, gained final acceptance by Oxford University Press only 7 days before Leopold's death, and the essays were published as *A Sand County Almanac* (1949). The final essay in that book, the essence of the essence, is "The Land Ethic," which, Leopold said, "sets forth, in more logical terms, some ideas whereby we dissenters rationalized our dissent, (1949:viii)." But the exact meaning of the land ethic remains controversial as interpreters of Leopold debate the metaphysical and moral commitments that relate to Leopold's mature conservation ethic. Environmental historian Joseph Petulla (1980:16, 204-205), for example, credits Leopold with advocating the moral equality of all species.[1] While the exact implications of bio-species equality remain unclear, this idea is difficult to interpret in a manner cordial to environmental management. Deep ecologists have generally questioned the right of humans to "manage" nature (Devall & Sessions, 1985).

Agricultural research has emphasized topics that promise to improve agricultural productivity. The framework of social values justifying this emphasis can be associated with Gifford Pinchot's approach to resource use. As the first U.S. Forester and the most influential resource manager in the U.S. Department of Agriculture during the first decade of this century, Pinchot adopted a utilitarian philosophy which em-phasized the development of resources to maximize the material well-being of all Americans.

Leopold began his career as a forester in the tradition of Pinchot, but was later forced to reconsider the materialistic and production-oriented aspects of Pinchot's approach. A careful study of Leopold's reasons for dissenting from Pinchot's approach provides insight regarding the weaknesses—from an environmental perspective—of a pure production orientation.

The purpose of this chapter is to examine Leopold's dissent and form tentative conclusions about a more environmentally sensitive ap-proach to agricultural research based on his central insights. In partic-ular, Leopold's evolving attitude toward environmental management is examined.

First, let me pose this question: Does the land ethic represent a final victory of Leopold-the-romantic over Leopold-the-manager? I believe that whatever the metaphysical and moral commitments of the land ethic, the ethic reflects substantive changes in Leopold's management practices. And these changes can be explained precisely in the vernacular of the state-of-the-art ecological theory of Leopold's day. Leopold never abandoned his intention to manage the environment, but he did criticize Pinchot's utilitarian model of management. He also developed his own theory. That theory and its applications represent the opera-tional meaning of the land ethic.

In the second part of the chapter, this operational meaning is illus-trated through Leopold's "paradigm" of failures in environmental management by examining three "case studies." On this basis, he proposed an integrated theory of environmental management, which I call "contextual" management.

In the final part of the chapter, I suggest that Leopold's ideas on management represent a significant anticipation of hierarchy theory, which is an important new approach to ecological theory (Allen & Starr, 1982; O'Neill, 1986). More importantly, this new theory has the potential to provide a more precise conceptual model for understanding environmental problems and their interrelationships, and showing how effective management is possible within environmentally determined constraints. Important implications of the hierarchical approach for agricultural research are discussed.

Leopold-the-Manager

In 1920, Leopold traveled from Albuquerque, New Mexico, where he was the U.S. Forest Service's director of operations for the Southwest Territories, to New York to attend the Sixth American Game Conference. In a report on game conditions in the Southwest, he enthusiastically described the predator eradication program he had initiated there. Leopold had formed a coalition of sportsmen and stockmen to eliminate wolves, mountain lions, and other large predators from Arizona and New Mexico: "But the last one must be caught before the job can be called fully successful," he said. He was concerned that "no plans for game refuges or regulation will get us anywhere unless these lions are cleaned out," but was optimistic about the future: "when they are cleaned out," he said, "the productiveness of our proposed refuges and plans for regulation of kill, will be very greatly increased."[2]

In 1943, Leopold repented his war on wolves in a graceful and humble little essay written explicitly for the collection (Meine, 1988). "Thinking Like a Mountain," Leopold's *mea culpa* on wolves, recounts his conversion from wolf eradicator to wolf protector. To understand Leopold's land ethic, we must compare Leopold's early game management strategies with the final point of his wolf policy.

Leopold's early wolf eradication policies were justifiable within the prevailing policy doctrines of the Forest Service, which expressed Pinchot's utilitarian approach to resource management. This is not surprising because Leopold obtained his master's degree in forestry at the Yale School of Forestry, which was established by a large gift from Pinchot's family. Immediately after graduation in 1909, when Pinchot was at the height of his influence over bureaucratic conservation policies, Leopold began working at the U.S. Forest Service. Pinchot had studied forestry in Europe, where sustainable forestry had a long history, and he was trying valiantly to apply the European model to the dissimilar conditions of the New World (Norton, in production, a). Building on his European experience, Pinchot placed his own, particularly utilitarian stamp on resource management practices. He explained (Pinchot, 1947:326) that his goal was to promote "the greatest happiness for the greatest number over the longest period of time " (Pinchot, 1947).

Pinchot emphasized the materialistic aspects of resource conservation, interpreting "happiness" as synonymous with material wealth. Although proud to be an outdoorsman, Pinchot did not let aesthetics play a role in his professional decisions (Norton, in production, a). A consummate politician and practical man, Pinchot easily accepted a market-driven, production-oriented approach to forest management. He measured success in forestry in board-feet of timber and in fees from the sale of grazing rights.[3] His slogan nods toward the future, "...over the longest period of time," but Pinchot's concern for the future was interpreted simply as management for sustainability of production for the foreseeable future. Pinchot's approach favored development, which would be constrained only if clear projections showed current policies would result in future shortages. These aspects of Pinchot's approach have been thoroughly discussed throughout the years (Hays, 1959).

Two further characteristics of Pinchot's scientific resource management eventually received Leopold's attention. Pinchot's approach was biased toward *atomistic* management practices and assumed an essentially *static* set of background conditions.

Scientific resource management was atomistic because it leaned toward single-species management. Foresters preferred monocultural plantings of a single species of fast-growing and marketable trees. Game managers emphasized deer production and, while production of pheasants and other birds might be encouraged as well, each game species had a "management plan" of its own. Pinchot's approach also was atomistic in a second and related aspect: Management goals were pursued individually. As early as 1923, Leopold questioned this feature of Forest Service policy:

> Take the Sapello watershed, which forms a major part of the GOS range in the Gila National Forest. Old settlers state that when they came to the country the Sapello was a beautiful trout stream lined with willows... but livestock has come in. And now the watercourse of the Sapello is a pile of boulders....

But Leopold did not criticize the management of grazing as such: "The remarkable thing about the Sapello is that it has not been overgrazed," he said. "The GOS range is pointed to with pride as a shining example of range conservation. The lesson is that under our peculiar Southwestern conditions, any grazing at all, no matter how moderated, is liable to overgraze and ruin the watercourses." From this example, Leopold formed a generalization: "The effect of unwise range use on the range industry, or of unwise farming on the land, are all too obvious to require illustration. What we need to appreciate is how abuses in one of these industries in one place may unwittingly injure another industry in another place " (Leopold, 1979).

So Leopold was quite aware of the problems caused by the atomistic bias of resource management practices prevalent in his day. However, Leopold did not fully recognize the revolutionary implications of this insight for resource management until more than a decade later. Only then did he develop an articulate explanation of its failings.

Leopold also came to realize that scientific resource management is based on an implicitly static model of biotic systems. It assumes these systems are regulated by a "balance of nature." While this idea was usually explained with pat examples of predator/prey equilibria, and was often explicitly qualified as overly simple, it still had a profound effect on resource managers such as Leopold. The idea encouraged them to understand stability as *resilience*. The resilience of a system is measured as the rapidity and accuracy with which a system, once disturbed, will regain its pre-disturbance equilibrium (Lewontin, 1969; Innis, 1974; Norton, 1987).

Any model which treats stability as simple resilience will necessarily be a static model—because return to the pre-disturbance state is considered normal. And because scientific resource managers see natural systems primarily in relation to their productivity, resilience was closely related in the manager's mind to sustainability of yields of one or more target species. As long as the system remains in equilibrium, one assumes it will produce a relatively consistent yield of the targeted species indefinitely. Or to put the point the other way around, once system stability is considered as resilience, stable yields of target species year after year are believed to be an adequate measure of the productive system's health.

Leopold's early approach to deer management can be understood as an application of this idea of health and stability. Because he believed in a natural equilibrium of the system that produced deer, Leopold considered wolves and mountain lions to be constraints that limited the annual production of deer for hunters. Removing wolves from the equation provided more deer for hunters, thus increasing the yield of deer to a maximum. Because hunters took the place of natural predators in controlling deer populations, a stable equilibrium could be maintained artificially. Indeed, Leopold apparently thought hunters were more controllable than other predators because of his comment in the New York speech, quoted previously, that once the lions were removed, "the productiveness of our proposed refuges and plans for regulation of kill, will be very greatly increased " (Meine, 1988).

At this time, Leopold was confident that variations in regulations on hunting (such as deciding annually whether or not to have an open season on does), provides the manager with sufficient tools to regulate deer productivity. By managing only to maximize and sustain deer yields, Leopold acted atomistically. He assumed he could hold the system in artificial equilibrium and he acted upon a static model. In this sense, Leopold applied Pinchot's atomistic and static management model early in his career.

Leopold's Transformation

Three examples, or case studies, of failures in resource management deeply affected Leopold in the late 1920s and early 1930s, causing him to reconsider the most basic elements of Pinchot's management model. Leopold traveled to Germany to inspect German forestry. Second, population irruptions of deer were occurring on the very game fields on which he had helped to remove predators. And, third, Leopold observed the dust bowl and tried, by guiding the loving attempts of his family to reclaim a burned out and eroded farm on the Wisconsin River, to redress the effects of farming that was too intensive for fragile soils.

Leopold had always been a thoughtful and self-critical manager, and his journals are sprinkled with worries about the dominant model he was applying, even as he applied it. For example, Leopold questioned the adequacy of "economic determinism" at least by 1923 (1979). As mentioned, his 1923 discussion of the Sapello watershed also questioned the atomistic feature of the dominant model. And Leopold even struck at the heart of the productivity-oriented model in the early- and mid-1920s when he developed and pushed through his proposal for the first wilderness area, the Gila, in the Gila and Datil National Forests. In defense of wilderness, he argued that the cultural advantages of retaining some relationship to our wilderness past outweigh and, in this case, justifiably limit the pursuit of material prosperity (Meine, 1988). But these dissents were mainly scattered observations registered as asides from the main task: to maximize productivity in the national forests.

In 1935, Leopold and five other U.S. foresters were invited by the Schurz Foundation of Germany to visit Germany to inspect and discuss new developments in German forestry. Leopold found the Germans, who had contributed significantly to Pinchot's productivity-based policies, in retreat from a century-long commitment to monocultural plantings of spruce. Superior yields of spruce from even-aged, single-species stands planted in the early 1800s had encouraged the Germans to apply this policy pervasively.

After two or three generations, however, the outstanding yields were not repeated. Leopold summarized (1936:374)

Litter failed to decay, piling upon the forest floor as a dry, sterile blanket which smothered all natural undergrowth, even moss. Roots ceased to penetrate the soil, lying in a tangled mat between the soil and the litter.... The topsoil developed excessive acidity, became bleached, and was separated from the subsoil by a dark band.

Leopold noted that these conditions became known as "soil-sickness." "In short," he concluded, "pure spruce, the precocious child of timber famine and 'wood factory' economics, grew into an unlovely and unproductive maturity."[4] The Germans, who Leopold credited with "teaching us to plant trees like cabbages," were retreating to a more natural approach, returning to mixed plantings, including native species.

By 1935, however, results of his deer management policies had already forced Leopold to undertake a fundamental reconsideration of conservation policy: Wolfless game ranges were prone to irruptions in deer populations. In 1924, Leopold left his position in the Southwest to accept an appointment at the Forest Products Laboratory of the U.S. Forest Service. That same year, biologists for the U.S. Biological Survey reported a runaway population of mule deer on the Kaibab Plateau. The range situation was critical, they said. Hunting, which had been halted on the Plateau when Roosevelt included it in the Grand Canyon Preserve in 1905, was reinstituted, but too late. Starvation caused a 60 percent reduction in the mule deer population during the hard winter of 1925-26. This example involved both predator eradication and hunting restrictions, so many debated whether one or both caused the phenomenon. From a distance, Leopold was puzzled.

He weakened his stance on predators, saying in 1925 that "it is important to avoid the extermination of predators, but there is no danger of this as yet," and he continued to refer to predators as "varmints." Then, in 1927, one of his colleagues in the Southwest began to issue warnings about a similar population explosion in the central, less accessible portions of Leopold's own Gila Wilderness (where hunting had been allowed and even encouraged). In 1929, the trend toward overpopulation, especially of does, was clear; Leopold then recommended that predator control efforts be shifted from mountain lions to coyotes. His former colleagues followed this advice but also, in a move that severely challenged Leopold's wilderness idea, began plans to build roads into the Gila to increase accessibility for hunters.

After a trip to the truly "wild" area surrounding Rio Gavilan in Mexico, Leopold concluded that reasonable and stable populations of deer were possible without removing predators. Leopold was significantly affected by this excursion, and later remarked that it was in Mexico that he "first clearly realized that land is an organism, that all my life I had seen only sick land, whereas here was a biota still in perfect aboriginal health. The term 'unspoiled wilderness' took on a new meaning," he said (Meine, 1988). Shortly after the trip, he asked fellow conservationists whether

> a normal complement of predators is not, at least in part, accountable for the absence of irruption (in the Gavilan)? If so, would not our rougher mountains be better off and might we not have more normalcy in our deer herds, if we let the wolves and lions come back in reasonable numbers? (quoted in Meine, 1988).

Once Leopold began to complete the collection of essays which became *A Sand County Almanac*, his student and friend, Albert Hochbaum, goaded him into including an admission that he "had once despoiled." Noting that they had just killed the last Lobo in Montana, Hochbaum said (quoted in Meine, 1988:453): "I think you'll have to admit you've got at least a drop of its blood on your hands." The resulting essay, "Thinking Like a Mountain," came to be the centerpiece of Leopold's mature conception of environmental management.

Leopold (1949:132-133) mused on his journey from predator eradicator to land ethicist. He told how he and his fellow crew members shot a she-wolf from a distance, an incident that occurred on Leopold's first "reconnaissance mission" as a forester. Leopold approached the mortally wounded she-wolf in time to see "a fierce green fire dying in her eyes." "I was young then, and full of the trigger-itch," he confessed. "I thought that because fewer wolves meant more deer, that no wolves would mean hunters' paradise." Leopold perceived the result: The bones of the desired deer herd, "dead of its own too-much," littered the mountain.

The theme of the essay is time: "Only the mountain has lived long enough to listen objectively to the howl of the wolf," he said. Thinking like a mountain is putting oneself in the time frame of the mountain, and "a mountain lives in mortal fear of its deer." And for good cause: "while a buck pulled down by wolves can be replaced in two or three years, a range pulled down by too many deer may fail of replacement in as many decades." Leopold was here recognizing how and why his deer management plan had gone astray: Utilitarian, economically calculating management was management conducted from the time perspective of humans. Consumed with human cares, we strive for "peace in our time," and find it hard to see things from the mountain's viewpoint, a viewpoint measured in ecological and geological time, not human time. (Leopold, 1949).

To help the reader understand the differing time frames, Leopold-the-romantic created a double metaphor. The mountain was personified: the dead sage became bones moldering, along with the deer bones, under the high-lined junipers. The mountain has not only a vegetative skeleton, however, it also thinks and feels fear. If we are to manage nature without causing havoc, we must think like the mountain thinks.

Another metaphor also is implied: the journey Leopold took from a "game manager" to a predator protector represents a symbol of the journey our culture must take in articulating and embracing a land ethic. And the metaphor of the living mountain drove the process; organicism helped him think like the mountain thinks.

At about the same time Leopold-the-romantic wrote "Thinking Like a Mountain," Leopold-the-manager was trying to initiate a state-funded study of wolf ecology. He also drafted a more technical article entitled, "The 1944 Game Situation." In it, he stated his conclusions in a scientific tongue,

> It is probably no accident that the near-extinction of the timber wolf and cougar was followed, in most of the big-game states, by a plague of excess deer and elk and the threatened extirpation of their winter browse foods... It is all very well, in theory, to say that guns will regulate the deer, but no state has ever succeeded in regulating its deer herd satisfactorily by guns alone. Open seasons are a crude instrument and usually kill either too many deer or too few. The wolf is by comparison, a precision instrument; he regulates not only the number, but the distribution, of deer. In thickly settled counties we cannot have wolves, but in parts of the north [of Wisconsin] we can and should. (quoted in Meine, 1988).

The pattern of Leopold's thinking about wolf management was reinforced by his more detached observation of the dust bowl phenomenon. Leopold was careful to clarify that he was not an agronomist, that his observations were only those of a nonprofessional. But Leopold had traced what he called "illness" in natural communities, especially in the Southwest Territories, even before speculating about a general theory of fragility in arid systems in 1923. He was moving toward an explanation of the dust bowl well before it occurred. His explanation was summarized (1949:132) in a succinct, but powerful passage near the end of "Thinking Like a Mountain." Just after summarizing the lesson he had learned about mountains fearing deer, Leopold made the comparison (1949:132):

> So also with cows. The cowman who cleans his range of wolves does not realize that he is taking over the wolf's job of trimming the herd to

fit the range. He has not learned to think like a mountain. Hence we have dust bowls, and rivers washing the future into the sea. (Leopold, 1949).

Leopold employed ecological terms to discuss his three case studies—German forestry, deer irruptions, and the dust bowl—and he thereby developed a paradigm of environmental management gone awry. Leopold was ready, tentatively at least, to articulate his criticisms of Pinchot-style management and propose his own broad "theory" of environmental management. The new theory represented an application of Charles Elton's communitarian model of ecological systems. Leopold met Elton in 1931 at a conference on natural cycles and they immediately became friends. Already acquainted with Elton's important book, *Animal Ecology* (1926), Leopold began to apply ecological theory to management problems in the 1930s (Meine, 1988). According to Elton's theory, species are understood functionally in terms of their contribution to a larger, biotic community.

Applying Elton's concepts, Leopold theorized as follows: The three case studies discussed all resulted from attempts to increase the output of a resource base through more intensive management. All three cases were considered initial successes because productivity increased dramatically. After a few years or decades, a crash in productivity occurred in all three situations. Elton's theory provided an explanation of the salient facts: Ecosystems embody redundancy, with many species fulfilling each function. The larger community changes and develops through time. This development is a result of external disturbances and local accidents affecting the community's loosely patterned successional tendency.

Elton's ecological model explained the breakdown encountered in managed systems: Complexity and redundancy, accumulated over millennia of evolution and competition, are reduced. The functions of the community—the flow of resources up and down the biotic pyramid—are simplified and reduced. The system is ill, but only sharp observers can see the gradual development of the illness. Then, usually in response to stresses like pest invasions, a hard winter, or a prolonged drought, the system collapses. Although it does not die entirely, it may re-equilibrate at a lower level of functioning. In the worst case, the system goes into a tailspin, as each downward turn in complexity results in lower productivity and vice versa, until the system is reduced to a desert.

Based on this scientific interpretation of the three events, Leopold introduced the concept of "fragility" as applied to ecosystems. He also drew on his early experience in the grazing office of the U.S. Forest

Service's Southwest District to broaden the concept of "carrying capacity" to apply to all types of stress to systems caused by human uses, including extractive uses. Leopold could now articulate and explain how the utilitarian approach to land management had failed. As we manipulate systems to produce maximum outputs of targeted species by interfering with normal ecological and evolutionary mechanisms, systems begin to lose their complexity. This impoverishment is not obvious because it occurs slowly and has a latent stage in which the redundancy (the internal complexity of the system) is only gradually eroded. Once the redundancy of function is removed, the system is vulnerable to breakdowns in its energy pathways, and the entire community can collapse. Elton's theory thus provided Leopold with a scientific vocabulary to explain the collapse of productive systems under prolonged stress and the latency of these collapses.

On this basis, Leopold developed what I call a "contextual" conception of environmental management on which each management problem and initiative should be considered on at least two levels: first as a *cell*, and second as a *cell-in-context*. On one level, the object of the management of deer, for example, is considered a complete system. But it is also a cell in a larger system. A cell in the system, Leopold learned, cannot be managed without also considering and keeping a watchful eye on the context (the vegetative cover) in which the cell is embedded.

Leopold rejected atomism when he emphasized that single species do not represent an adequate cell for management purposes. He further departed from atomistic management when he argued that cells must be managed by considering the entire range of human purposes. And he insisted that any successful management model must conceive stability dynamically. All of these insights can be summarized by saying that Leopold insisted that cell management always occurs in a larger "context," or in a larger functioning system.

At first, Leopold thought deer-wolf-hunters was a manageable cell in the system; but he learned that under the conditions in the Southwest, it was not. He learned to pay attention to "the mountain," the vegetative cover that gives structure, complexity, and a certain type of stability to ecological communities. Leopold, building on Elton's theoretical conception of a community, concluded that we must always manage any species *in its context*, i.e., as a member of an ecological community. That management is not considered adequate if the activity decreases the overall complexity of the contextual system (causes "illness" in the community); or if effects of the activity disrupt productivity in a larger context (such as when grazing destroys watercourses and trout fishing). Under these conditions, the larger system is in danger of destabilization (such as on the deer-infested mountain).

Leopold's scientific, contextual model implied a two-step process for forming environmental policy: (1) managing cells according to maximization criteria is permissible, provided the maximization plan also passes a broader criterion; and (2) economically motivated activities must not threaten the "health" of the larger context. In this respect, Leopold's approach seems very similar to that developed more recently by economist Talbot Page. Page (1977) argues that a wise materials policy is first guided by a conservation criterion, which delimits a set of ecologically acceptable policies. A second criterion, which further distinguishes the policy options that pass the conservation criterion, chooses the policy that is the most economically efficient.

Leopold, like Page, insisted that policies unacceptable for conservation reasons be eliminated from further consideration. Page and Leopold also agree that the system of analysis cannot be simply aggregative. The operation of economic analyses is limited to particular, productive subsystems. Analyses of productivity data within cells is not aggregated with data concerning the health of the larger context. Unlike Page, who applies this reasoning only to a "materials policy," Leopold applied similar reasoning to all management problems. Leopold also argued that the conservation criterion should be ecologically informed and conceived on an organic model, including a strongly normative conception of "ecological health." This model should be capable of determining which activities contribute to illness in the environmental context (Page, 1977).

Hierarchy Theory

Despite important theoretical breakthroughs, Leopold faced an impasse; his theory told him to manage contextually, taking into account the varied scales of time and space that interlock ecological communities. "Ecology remains the fusion point of the sciences," he said (1939:727). And he gave credit to "economic biologists" for helping promote the field of ecology. But ecology was not equal to its task, he feared:

> The emergence of ecology has placed the economic biologist in a peculiar dilemma: with one hand he points out the accumulated findings of his search for utility, or lack of utility, in this or that species; with the other he lifts the veil from a biota so complex, so conditioned by interwoven cooperations and competitions, that no man can say where utility begins or ends.

This extremely important essay represents a summary of what Leopold had learned about environmental management and explicitly refers to all three of the "paradigms" discussed in this chapter.

Leopold's elegant theory unfortunately was impossible to apply, because of the state of development of ecological theory in his day. Two weaknesses in the communitarian model prohibited Leopold from applying the theory more precisely. First, the communitarian model suffers from a difficult problem of "parts and wholes." What counts as a unit for ecological study? What counts as a unit for efficient management? Even in the 1930s, ecologists were interested in energy flows as an organizing concept for ecological systems. And, in some cases, ecological systems seem quite naturally bounded, as in the case of a spring-fed lake with no outlets. But even in these optimal cases, ecologists must caution that their ecosystem "boundaries" are at best permeable membranes; they are more accurately viewed as temporary conveniences, not as actual, closed systems. Most cases in nature are infinitely more complex than lakes without outlets—habitats and microhabitats usually intertwine in an overlapping complexity that cannot be unraveled.

Second, while communitarian ecology provided a general framework for conceiving species and complexes of species as embodied in a larger, contextual community, it lacked an adequate conception of contextual stability. But Leopold's contextual theory required a dynamic conception of stability and ecosystem health, one capable of relating a resource-producing subsystem to its larger context because the context changes and develops as well. Stability conceived as resilience of the system, as it is often conceived even today, is too static to adapt to Leopold's new model. Management must be implemented taking into consideration the larger context, which is expected to change and develop in its own time frame. Inadequate management of a cell can cause destabilization of the context.

As Leopold emphasized in "Thinking Like a Mountain," these interlocking systems embody differing scales of time—the mountain must think more slowly than the hunter or the deer, and the environmental manager must think like the mountain in order to manage deer/hunter interactions in their larger context. Stability and health in nature must be measured in ratios, by comparing rates of change. Because it lacked a serviceable conception of dynamic stability of larger systems in which management units are embedded, Leopold's management theory was disabled.

He recognized the problem clearly and argued persuasively that the equilibrium model was inadequate for the complexities of management. He warned that a static conception of the balance of nature cannot successfully model natural systems.

> To the ecological mind, balance of nature has merits and also defects. Its merits are that it conceives of a collective total, that it imputes some utility to all species, and that it implies oscillations when balance is disturbed. Its defects are that there is only one point at which balance occurs, and that balance is normally static (Leopold, 1939).

Leopold launched immediately into a discussion of ecological systems as energy pyramids and argued that "Each species, including ourselves, is a link in many food chains." Leopold then noted that "the trend of evolution is to elaborate the biota," to make it more complex and to multiply the channels through food chains by which energy flows to the top of the pyramid. Contextual management conceives the context of management cells to be the larger ecosystem in which the cell is embedded—and that context is inevitably simplified by management activities. It remains healthy only if the energy flows upward in the pyramid. Leopold described this elaborated biota as "a tangle of chains so complex as to seem disorderly, but when carefully examined the tangle is seen to be a highly organized structure. Its functioning depends on the cooperation and competition of all its diverse links" (Leopold, 1939: 727-728).

Leopold, stymied in his attempt to develop positive prescriptions based on ecology, did the next best thing. He fell back on his governing organicist metaphor and an analogical implication from medicine: If no cure exists for the disease, practice preventive medicine. Caution and humility should, he concluded, guide environmental managers.

Recognizing that humans are a part of the system they manage, Leopold introduced a crucial distinction based on the "violence, rapidity, and scope" of human changes. Human changes differ from evolutionary changes, he said, which "are usually slow and local." Leopold therefore distinguished between management activities that are sufficiently violent to disturb their context from those that are not—the former interrupt the energy flows of the community in which they are embedded, and the latter do not. "The combined evidence of history and ecology seems to support one general deduction," he said, "the less violent the man-made changes, the greater the probability of successful readjustment in the pyramid." (Leopold, 1939)

Leopold was therefore struggling toward a dynamic conception of ecological health and stability. His tentative model emphasized complexity—the ecological processes that allow ecosystems to perpetuate themselves, rather than diversity itself. By introducing organicism, Leopold was proposing a new conceptualization of the goals of contextual management: to protect the autogenic functioning—the self-determination—of the larger system. Economically motivated management is acceptable until trends in the exploited subsystems start to accelerate changes in the larger, slower-moving system. A system that is rapidly losing complexity as a result of pervasive and violent changes in its subsystems is ill. But ecology failed Leopold in this aspect because the fledgling science had provided no quantitative measure, or even a clear qualitative conception, of autogenic functioning.[5]

Despite difficulties of application, Leopold's central and correlative concepts of ecological fragility and ecological health were highly developed scientific constructs. He developed three parameters relating these concepts to human activities based on various levels of violence. The extent of the simplification caused by human changes in ecosystem functioning varies according to the following three conditions:

- **The fragility condition:** Every natural system has a certain degree of integrity, or self-determination, resulting from the accommodation that the indigenous species have made with the land and among themselves. Some systems, such as those in the arid Southwest Territories, are extremely fragile ("set on hair trigger"), and are extremely susceptible to deterioration in energy circuits as a result of overgrazing or other pervasive disturbances. They are prone to illness.
- **The population condition:** Whatever the inherent integrity/fragility of a given system, larger human populations result in more pervasive changes and tend to cause greater modifications and simplifications ("greater violence").[6]
- **The time scale condition:** Natural systems have evolved in evolutionary and ecological scales of time, in which changes occur more slowly than human changes (We do not yet "think like a mountain"). The more rapid human-induced changes, the more likely they are to destabilize the complex systems of nature (Leopold, 1939).

Leopold never questioned the *right* of humans to manage; he questioned, rather, our *ability* to manage, arguing that given the present state of ecological knowledge and theory, we often fail to foresee important effects of management initiatives. Since Leopold's day, little progress has been made in developing a dynamic conception of stability and, until recently, not much more has been learned about how to relate processes occurring in different time frames.

Recently, however, a new and highly promising theoretical approach, "hierarchy theory," has emerged within the field of ecology. This theory significantly resembles Leopold's communitarian model and contextualist approach to management. The new approach has the potential to further clarify the concepts that plagued Leopold's theory of environmental management. Hierarchy theory is based on general systems theory and focuses more on the complexity and internal organization of systems—what Leopold called "integrity" (Leopold, 1949)—and not as much on their diversity.

According to hierarchy theorists, natural systems exhibit complexity because they embody processes that occur at different rates of speed. In general, larger systems (such as a community) change more slowly than the microhabitats and individual organisms that compose them, just as the organism changes more slowly than the cells that compose it. Furthermore, the community survives after individuals die and, while changes in the community affect (constrain) the activities of the individuals in it, the individuals themselves are unlikely to affect the larger system. This occurs because the individual is likely to die before the slow-changing system in which it is embedded is significantly altered by its activities.

This is not to say that elements have no impact on systems that provide their context. The elements, often called "holons" in the context of the analysis, are "two-faced." Each holon "has dual tendency to preserve and assert its individuality as a quasi-autonomous whole, and to function as an integrated part (of an existing or evolving) larger whole (Koestler, 1967; Allen & Starr, 1982)." As a part, the holon affects the whole, but scale is very important—the "choices" of one element do not significantly alter the whole, but if that part's activities represent a trend among its peers, then the larger, slower-changing system reflects these changes on its larger and slower scale. One cell turning malignant does not affect an organism significantly unless a trend toward malignancy is thereby instituted. If such a trend is instituted, then the organism might eventually be destroyed.

This multi-scale approach to time and space in ecology is reminiscent of Leopold's metaphorical discussion of differing scales of time and our perception of them in "Thinking Like a Mountain." Hierarchy theory provides a more precise conceptual model for what Leopold called "the land," which was a slower-changing system composed of many faster-changing parts. He explicitly commented that our failure to see deterioration in the land community results from our failure to recognize that ecological and evolutionary changes occur in a slower scale of time than the scale perceived by humans. Agriculturalists and game managers focus on the rapid-change systems that produce annual crops. The mountain, as Leopold explained metaphorically, must consider the value of wolves in a broader perspective (Leopold, 1949).

When nature is viewed hierarchically, as a system of parts embedded in larger and larger wholes, holons that are higher in the hierarchy represent the environment of lower holons and constrain their activities (Allen & Starr, 1982). Allen & Starr argue that "the positive aspects of organization emanate from the freedom that comes with constraint. The constraint gives freedom from an unmanageable set of choices; regulation [from above] gives freedom within the law." According to hierarchy theory, then, natural systems are organized units embedded in a hierarchy, with larger, slower-changing systems (the environment, or "context") that determine the range of choices available to the smaller systems. This abstract, hierarchical model can be applied at many levels—cell/organ, organ/organism, organism/microhabitat, microhabitat/ecosystem, ecosystem/bioregion, etc. It provides a sliding scale of concepts for analyzing relationships among the parts and wholes of living systems.

In ecology, which emphasizes the interrelationships among systematic elements, hierarchy theory provides a useful tool for analyzing the multi-layered complexity of natural systems, and can potentially model the dynamic relationships among their parts. The hierarchical model may also begin to identify a managerially useful concept of dynamic stability and ecosystem health. It can locate this concept in the interrelated functions of faster-and slower-changing systems. Accelerating changes in normally slow-changing systems may indicate deterioration or illness in the land community.

Leopold emphasized an aspect of the hierarchical model that has not, as yet, received much attention by hierarchy theorists who mainly understand systems by tracing the constraints that higher-level holons have on their parts.[7] One might say that, whereas hierarchy theorists view the hierarchy from above, Leopold sees the same hierarchy from below. Leopold, as a manager, was interested in the effects that the parts have upon their context, when altered violently and pervasively by modern civilizations with modern technologies. Leopold's observations in the Southwest convinced him that, if human resource users are considered the parts and the land system is considered the whole, modern methods of resource use were causing swift changes in that larger system.

Allen and Starr (1982) briefly discuss these large-scale matters affecting whole landscapes, emphasizing that changes recognized at the individual level are normally "damped out" in the higher-level systems. They note that the dust bowl represents an example in which the activities of individual farmers had little effect on the large-scale environment of the plains states as long as they occurred on a limited scale. Eventually, however, as more and more land was converted to monocultural agriculture, a threshold was reached and the plains system, which had been changing slowly for millennia, experienced rapid change as a result of a pervasive trend in the behavior of individual farmers.

This explanation corresponds to Leopold's: The dust bowl resulted from intensive and pervasive application of monocultures and intense grazing in a bioregion that was ill-suited to intense use because of its cyclical patterns of rainfall and aridity. Similarly, Leopold, as a game manager early in his career, saw deer-wolves-hunters as a cell that could be managed, and he destroyed wolves to maximize deer for hunters. However, experience taught him that a cell must be considered a holon, and it therefore exists also as a part of its larger environment. The removal of wolves caused an irruption of deer, and deer overgrazed their browse, exceeding the capacity of the larger system to equilibrate these changes. A parameter (or constraint) is exceeded; the larger system becomes unstable, because redundancy is lost, and under stress, rapid change such as dust storms and even desertification become more likely.

If a pervasive trend exists in environmental management today, it is toward a more holistic, or at least more contextual, management model. For example, in the "Foreword" to its brochure on Chesapeake Bay management, the Environmental Protection Agency notes: "The Bay is, in many ways, like an incredibly complex living organism. Each of its parts is related to its other parts in a web of dependencies and support systems. For us to manage the Bay well, we must first understand how it functions" (EPA, 1982).[8]

Hierarchy theory again raises the question of parts and wholes. How seriously should we take any particular designation of "parts" and "wholes," or "cells" and "contexts"? Hierarchy theorists (Allen & Starr, 1982:10) avoid this problem by insisting that their model resides not in reality, but in our understanding of reality (the model is "epistemological," not "metaphysical") (Allen & Starr, 1982). Whether natural systems are themselves hierarchical is not the point; viewing systems as hierarchical makes their complexity intelligible to us. Any designation of parts and wholes is therefore a temporary expedient for a particular analysis, a result of choosing a particular "observation set," which is "a particular way of viewing the natural world." (O'Neill, 1986; Allen & Starr, 1982). This approach avoids commitment about which aspects are fundamental: "Since [no viewpoint] encompasses all possible observations, neither can be considered to be more fundamental. When studying a specific problem, the scientist must always focus on a single observation set. However, when developing theory, many observation sets must be considered." (O'Neill et al., 1986)

But O'Neill and his co-authors also recognize that "the task of choosing an appropriate system for investigating a particular phenomenon is inseparable from consideration of underlying organization and complexity." Applications of hierarchy theory to environmental management, therefore, focus on the organizational complexity of the particular systems to be managed.

Whereas hierarchy theorists are seeking an abstract model that can be applied at all levels of biological organization, the scale of concern to the manager will be indicated by the problem at hand. Consequently, based on the philosophy of hierarchy theorists that freedom and creativity develop within constraints, we can expect that hierarchy theory may find more immediate application to situations constrained by management concerns. Intensive efforts to apply hierarchy theory to complex ecological systems under stress may therefore encourage further elaboration of hierarchy theory itself.

Hierarchy theory raises especially interesting prospects for the future of agricultural research. Conversion of natural systems to agricultural systems involves the merger of two hierarchically organized systems—a human, social system and the evolutionary system of nature. The problem is to find a productive and reasonably stable compromise—one that respects the economic requirements of agricultural production, and the aesthetic and lifestyle requirements of the local culture, while protecting the complexity of the larger context.

Again, Chesapeake Bay management provides an example. Governments in the region are firmly committed to holistic management of the bay ecosystem. That is, they now accept that the specific resource-producing systems surrounding the bay must be managed without exceeding parameters imposed by the health of the larger bay system. Agriculture, fishing, recreation, tourism, etc., can be conceived as socially driven systems that are partially parasitic on the larger, slower-changing bay system.

The most acute current problem in managing the bay—the algal blooms resulting from too much run-off from yards, fields, and pastures—can then be interpreted as rapid change in a normally slow-changing system. This change, according to the model, results from accelerated changes in residential development patterns, agricultural production, and consumer preferences. These systematic changes result from social, economic, and demographic trends, such as rapid residential development, increased fertilizer use, and increased demand for seafood. Farmers, homebuilders, and fishermen tend to make decisions according to short time frames, seeking, as Leopold said, "peace in our time."

As a result of these changes, the bay is undergoing accelerated change itself, perhaps the most threatening of which is a significant reduction, over the last decade or so, of the highly productive submerged grass beds. By applying the concepts of hierarchy theory, modeling dynamic stability as a relationship between the various subsystems of the watershed may be possible. The goal, in this case, would be to achieve a relationship in which the bay's subsystems function and develop without disturbing the rate or trajectory of the slower changes in the larger context.[9] When changes in the larger system result from unprecedented changes in its parts, the larger system is being destabilized. Its processes and equilibria which were for centuries able to suppress changes in productivity of component subsystems, are now being overwhelmed by rapid subsystem changes.

Public officials have launched an unprecedented regional effort to manage the bay. Can they reverse the illness in the larger, contextual system, on which so many highly productive activities depend? That depends on the accomplishment of three tasks: (1) carefully monitoring trends in the component subsystems, including agriculture, to determine whether these trends are likely to have major effects on the system; (2) developing the scientific ability to monitor the bay system and model changes in it to distinguish destabilizing trends from cyclical and other autogenic changes in the larger system; and (3) achieving a grassroots political consensus that restrictions on activities in the component systems are necessary and in the best interest of everyone. Each of these three tasks represents a major challenge, but the challenge must be met for the health of Chesapeake Bay to be protected.

Leopold on Management

At the beginning of this chapter I asked whether Leopold-the-romantic won out over Leopold-the-manager. I believe that Leopold never lost his resolve to manage natural systems. He did, however, reject the Pinchot-style approach to management, which was dominant in his day, because that style was atomistic, and because it falsely assumed that natural systems can be modeled statically. With Leopold's contextual theory of management, policy would not be determined only by human economics. Economic considerations would be constrained by a concern for the health of the larger system in which economic activities occur.

Leopold believed that application of the organicist, rather than the mechanistic analogy (trying to "think like a mountain"), helps us visualize the management problem. Lacking precise measurements to relate the changing components to the larger whole, however, Leopold advocated caution and loving respect. In his most explicit philosophical discussion of organicism, Leopold (1979) said he "felt intuitively" that the earth was more than a mere mechanism, but that language is inadequate to resolve such questions.

In spite of these skeptical concerns, Leopold-the-romantic longed for a day when his fellow Americans would adopt an attitude of love and respect for the land and embrace the organic metaphor. Whether or not Leopold should be attributed to a literal belief in organicism, or a full-scale commitment to the Gaia hypothesis, remains an interesting historical and philosophical question. However, this question does not significantly affect his management theory. All that is necessary for this theory is recognition of the *usefulness* of the organic metaphor to illuminate the complexities of cell-management within a larger, changing context.

Throughout his life, Leopold believed in the necessity and desirability of environmental management. Leopold-the-romantic never overwhelmed, but only tempered Leopold-the-manager. In the end, the two were united in a single persona—Leopold-the-land-ethicist.

Implications for Agricultural Research

In conclusion, I want to summarize several implications of Leopold's dissent from the production-maximizing approach to "management." From these suggestions, we can extract new approaches to agricultural research that are more sensitive to environmental limitations.

First, environmentalists have often voiced concern about the tendency, often exhibited in the production orientation, to analyze policy in purely aggregative terms. Leopold's contextual approach and hierarchy theory explain why environmentalists are wary of purely aggregative analyses. Because systems change at different rates of speed and in different scales of time, aggregating data about cells with data about the larger context is not possible.

Risk analysis, for example, attempts to achieve a single-value, unidimensional characterization of all values, considering increments of risk as costs and decrements of risk as benefits. Hierarchy theory, on the other hand, presents constraints from outside the system of production, based in natural systems themselves. These constraints cannot be reduced to costs and benefits within the system, and therefore cannot be aggregated with other costs and benefits.

Second, contextualism and hierarchy theory provide a conceptualization for talking about limits, even while maintaining an important role for productivity. A hierarchical system allows the analyst to emphasize productivity that respects the constraints inherent in larger natural systems. Hierarchical thinking suggests that maximization is permissible as long as parameters (or constraints) of the larger system are not exceeded.

These constraints are not based on metaphysics or controversial moral assumptions, but on a scientifically based *model* of "land health." It assumes, therefore, a biological and climatic criterion for judging the acceptability of agricultural activities based on Leopold's three parameters of limitation: the fragility condition, the population condition, and the time scale condition.

Third, farmers who are near bankruptcy have an incentive to abandon the production-maximizing approach—what they lack are the means to maintain profit margins while reducing inputs. For example, they often feel "addicted" to chemical fertilizers and large machinery because of the emphasis on productivity.

Agricultural research should provide alternative approaches, based on an analysis of local conditions, constraints, and opportunities, to empower farmers to reduce costs and maintain profit margins without maximizing production. Chapter 5 of this volume, by Miguel Altieri, provides a model for developing locally appropriate alternatives to high-cost, environmentally destructive methods.

Finally, the proposed hierarchical conception of agricultural resources management suggests three new objectives for agricultural research:

- Examine the limits inherent in particular, local systems to seek methods of expanding productivity without exceeding these limits.
- Determine methods for specific regions that can lower input costs and allow farmers to increase profit margins without increasing production.
- Assess technologies to determine which are merely "addictive" (so costly that they require the farmer to increase productivity to remain competitive) and which can empower farmers by offering alternative production options.

Notes

1. Also see John Passmore, (1973) Ch 5. Passmore suggests that Leopold was committed to the rights of nonhuman species and implies that his views of nature are "mystical." For an extreme statement of the view that nonhuman elements of nature have inalienable rights, see Nash. Also see Norton (1986).

2. For an excellent treatment of Leopold's changing approaches to deer and wolf management, see Susan Flader's *Thinking Like A Mountain;* 1974. Quotations are taken from Meine (1988) who reproduces portions of Leopold's 1920 speech, "The Game Situation in the Southwest," on p. 181.

3. For a useful discussion of the production-orientation of early conservationism, see Hays, 1959.

4. Also quoted in Meine (1988) in 353-354.

5. The most significant developments since Leopold's death have involved applications of information theory and systems theory to ecological systems. See Ramon Margalef, *Perspectives in Ecological Theory* (1963).

6. See George Woodwell, "On the Limits of Nature," in R. Repetto, Ed., *The Global Possible* (Washington, DC: World Resources, Inc., 1985).

7. See Pattee (1973) for a brief discussion of effects flowing upward in a hierarchy.

8. For another example of contextual approaches to management that is applied to protecting old growth and biodiversity throughout the entire region of the Pacific Northwest, see Larry Harris (1984).

9. See Allen and Starr (1982) Chs. 3, 8, and 9 for a discussion of scales and "filters," which represent conversion functions between systems operating on differing scales.

7

Ethics of U.S. Agricultural Research in International Development

Merle L. Esmay

Introduction

The *Beyond the Large Farm* project has formulated two main objectives. The first is a philosophical examination of agricultural research in light of the various social and moral values it might be expected to serve. The second is to clarify the range of options available to an individual scientist working within the USDA/land-grant system, and to determine how (and whether) a broader set of social goals might influence research methodology at the micro level.

This chapter extends beyond the domestic land-grant system to review the effect U.S. agricultural research and education may be having on international development.

The International Problem

The primary problem is determining how to redirect agricultural research from its traditional singular short-term focus on increased productivity to a broader long-term concern with sustained agriculture, including preservation of natural resources and improvement of quality of life. L. Brown (1989) states that the latest annual physical examination of the earth's vital signs indicates that "The forests are shrinking, its deserts expanding, and its soils eroding—all at record rates. The protective ozone layer is being depleted. The very temperature of the earth appears to be rising, posing a threat of unknown dimensions to virtually all the life-support systems on which humanity depends."

Brown goes on to say that for the 1990s to be a turnaround decade for ensuring a habitable planet for future generations depends on developing new energy strategies, stabilizing population, and reducing both land degradation and development-throttling Third World debt.

Robert Rodale (1988) says that the lack of sustainability of agriculture is the problem. He goes on to explain that about 10,000 years ago, agriculture began. For the first 99 centuries of this period, farmers produced crops using nothing more than the internal resource of their farms, themselves, and their families. There were no inputs in the modern sense. Farming was not industrialized. Outside inputs such as machinery, tractor fuel, pesticides, and manufactured or mined fertilizers were not a part of the method of production.

The question of appropriateness of technology and sustainability of agriculture arose more quickly in the developing world according to Francis (1988). Mindful of the success achieved in agriculture in the temperate zone, many countries in the Third World have strived to emulate this industrial model, often with little impact. Lack of a domestic source of petroleum, or competitions for petroleum from other sectors, and an escalating international price for this resource have put current technology out of reach for many who farm the land. Francis quotes Agricultural Economist Vernon Ruttan of the University of Minnesota as stating bluntly that "Agriculture in many countries of the developing world has not yet entered the fossil-fuel age. For many reasons, some countries in the world may never reach that stage of development." They are being forced to seek another type of technology—one based on renewable resources—for which few advanced models are available.

Bernard D. Knezek et al. (1988) advocates a new 20/20 vision for agriculture. That is, reducing current high-energy, agricultural inputs to 20 percent. This is a drastic challenge, and one of the first times that a crops and soils professional (a producer) has admitted that some reduction in yields may have to be considered and accepted. He goes on to state that answers will require a systems approach to research rather than the traditional disciplinary approach. Traditionally, scientists operate independently in designing and conducting research projects on a single production component and then publish the results in disciplinary journals.

The idea that system science must be used for agricultural sustainability is also emphasized by Rawlins (1988):

> Look at nearly any large system and you will find a tragic waste of resources resulting from the failure to develop holistic management strategies that provide the necessary linkage between the systems com-

ponents. Typically, we make isolated decisions at the local level, allowing the larger system simply to evolve as an aggregation of local decisions. Maximizing the benefit for an entire system will require different decisions from those that result through maximizing the return from each component.

B.A. Stout and R.F. Nehring (1988), predict that "Rising United States reliance on imported oil and other aspects of world oil politics make the probability of another energy crisis in the 1990s appear rather high. United States agriculture could produce enough energy from biomass to meet its needs and possibly to supply other sectors of the economy, but only if energy prices rise substantially above 1988 levels."

Goals and Definitions

This chapter presents and analyzes past and present trends in international agricultural research and education for the purpose of formulating some guidelines and making recommendations for the future. Specific attention is devoted to the discussion of research directed toward sustainable agriculture as opposed to research which is narrowly focused on production. Sustainable agriculture, farming systems, and cropping systems are defined as relevant to developing counties. The systems approach is reviewed as a method to formulate a future direction for agricultural research and education. Generalized agricultural education vs. specialized education are discussed and analyzed as they apply to the training and education of international graduate students.

In January 1989, *The Agronomy News* reported a consensus achieved by 350 members of the ASA, CSSA, and SSSA following a "freewheeling, two-hour discussion": "A sustainable agriculture is one that, over the long-term enhances environmental quality and the resource base on which agriculture depends; provides for basic human food and fiber needs, is economically viable, and enhances the quality of life for farmers and society as a whole." Francis and Hildebrand (1989) define a close and complementary relationship between sustainable agriculture and farming systems research and extension. "Sustainable Agriculture is a set of goals or objectives for agricultural systems. Farming systems research and extension is a methodology that incorporates a systems, or holistic perspective." The shortened "farming systems" was expanded to "farming systems research and extension" to emphasize the necessity of the communications linkage at the farm level, for which extension must play a primary role.

The systems approach is holistic, but without the structured methodology that has been developed for various farming systems programs. The systems approach entails competent professionals in appropriate disciplines analyzing a problem, considering all relevant components. The systems approach lends itself to computer modeling so real-world conditions can be represented as accurately as possible, and various "what-if" questions can be studied for management purposes and decision-making. An outgrowth of the farming systems research and education group in Florida has been the formulation of an ad hoc task force for the development of an Association for Sustainable Farming Systems. The task force is chaired by Dr. George H. Axinn (FAO Representative, India and Bhutan, P.O. Box 3088, New Delhi 110 003 INDIA). Farming systems has lost its appeal as a "buzzword" in the international programs and projects arena, thus an attempt is underway to relate farming systems with sustainable agriculture.

Regional Research Institutes

The present network of regional research institutes was developed cooperatively by the Ford and Rockefeller Foundations beginning in the late 1950s. They were planned to be commodity (crop) oriented. Each institute was located in a specific geographic region and focused on the one or more critical crops of that region. The International Rice Research Institute (IRRI), located in the Philippines, is a primary example where rice culture is the main thrust for the Asian countries.

The term "farming systems" was in use at the International Crop Research Institute for Semi-Arid Tropics (ICRISAT) in India as early as 1977. Scientists had devised a unique wide-ridge and furrow system (two-meter-wide ridges) for the three or so semi-arid crops for which the Institute was responsible. The wide ridge would accommodate two, three, or four rows depending on the crop, and all animal draft and wheeled machinery would be confined to the furrow. Detrimental soil compaction was therefore confined to the furrows, and irrigation water could also be applied through the furrows. Farming systems was later defined more broadly to include the entire farming operation (livestock, management, marketing, quality of life, etc.), and the ICRISAT programs were more correctly called "cropping systems."

Maximizing Production

Historically, the early approach of agronomic research at the international research institutes was to maximize yields through breeding and nutrition under ideal controlled conditions. This was undoubtedly a carry over from U.S. and Western World trained scientists. The results were generally that only well-to-do farmers had the capital and capability of providing the expensive inputs, and thus could take the additional risk of trying new innovations. In the early 1960s, high-yielding rice varieties were developed by IRRI in the Philippines (IR-8 was the first well-known variety released). Large farmers realized substantial increases in yields, but small farmers were mainly excluded. As a result of the dominant concentration by plant scientists on maximizing yields, other critical factors were overlooked. A difference in taste developed which the natives did not like. Also, the rice shattered easier at much higher moisture contents and resulted in higher field shattering losses, making new harvesting, handling, and drying methods necessary.

Outreach and Industrial Extension

The early lesson of IRRI and other international research institutes was that strong linkages are needed between regional countries and growers. This lesson resulted in placing emphasis on the development of crop varieties that would increase yields under conditions more typical of small farmers.

In the field of agricultural engineering, the outreach program became known as industrial extension. Dr. Amir Khan, who headed the IRRI Agricultural Engineering Department for more than 10 years during the late 1960s and early 1970s, first recognized that the development of many fine prototype machines at IRRI had no impact or value if they were not accepted and adopted by local manufacturers and used by local farmers. An extension program was formulated to work with the small manufacturers (Khan, 1985). They needed significant guidance and training because few had any formal engineering background. To follow through with his approach and strengthen the outreach program, Dr. Khan accepted an assignment in Pakistan for 5 years (1978-1982). He worked directly with the outreach program for manufacturers and farmers. This pattern of outreach to local manufacturers has become well-known as industrial extension and is being used in many developing countries (Khan, 1979).

National Agricultural Research Institutes

Most developing countries have justifiably established their own agricultural research and development (R&D) institutes. These are generally organized directly under the Ministry of Agriculture. For the most effectiveness, these institutes should be closely affiliated with all other in-country research programs (such as universities), and more broadly to out-of-country regional research institutes. Some countries have been more successful than others, depending on resources, trained personnel, and management. Many American agricultural research scientists and institutions have been involved in the research programs and projects of these national R&D institutes. In some cases, collaborative research agreements (contracts) have been made between the national institutes and American institutions (sometimes land-grant universities). American scientists have served as consultants and many of the national scientists have been trained at American universities. In addition, financial support from American agencies and foundations has been provided for many of the national agricultural institutes. Thus, American agricultural research methods and approaches have significantly influenced these national institutes in many countries.

Adoption and Popularization

The major problem with the national agricultural engineering (mechanization) institutes has been that the research results are not readily adopted and popularized by the in-country manufacturers and farmers. The problem is two-fold: The research institutes have a misguided research thrust, as well as an ineffective working relationship with farmers, manufacturers, and extension programs.

Dr. Khan emphasized in a presentation at the 1988 National Summer American Society of Agricultural Engineers (ASAE) meeting that researchers and the institutes tend to be independent regarding what research projects they choose to undertake. The selection of projects is overly dominated by top-down orders and personal interests of the researchers. Therefore, the lack of concern for and awareness of the critical problems of the local agriculture is a problem. Researchers often find that working on creative, high-tech projects, for which the results are more publishable in international professional journals, is more exciting and professionally beneficial. They have generally been trained in a specialized area (such as their graduate program dissertation topic abroad), and find it more comfortable and possibly more satisfying to pursue research in that specialty area. For researchers to make their own selection of creative and interesting projects (or follow top-down orders) is much simpler than determining current, real-world problems and working closely with farmers and manufacturers to provide appropriate and acceptable solutions.

A 2-year-old FAO/UNDP research and development mechanization project was underway in Thailand in the early 1980s (Esmay, 1983). The objective of the project was to develop a more appropriate and adaptable animal-drawn plow for upland conditions. Farmers were particularly interested in a plow for single animal draft, which is common in China, Taiwan, and South Asian countries. The first year was spent organizing the project and obtaining prototype models of plows from many countries, and the second year in testing, analyzing, and modifying selected plows at the research station. The selection of prototypes from all possible sources was a good approach. Many developing countries believe borrowing technology from other countries is degrading so they try to create their own. Japan, Taiwan, and South Korea have certainly profited by borrowing from all available sources.

The second year of the project was essentially wasted because it was spent testing and monitoring selected plows under controlled research station conditions without any contact with farmers and manufacturers. The tendency of researchers is to develop in isolation what they believe to be the best plow or product and then take it to the farmers and manufacturers. Often, however, some critical concerns or problems of the farmers or manufacturers have been overlooked and make the new technology unacceptable. Early working relationships and trials by farmers would eliminate wasted time and effort and provide an acceptable and appropriate result much sooner.

The appropriate solution (the plow, as in the Thailand project) must also be capable of production by local manufacturers and shops. This requires close interaction and a follow-up industrial extension program. The local artisans need to be upgraded in skills, as well as quality control, management, and marketing.

Educational Problems and Solutions

Undergraduate Specialization Vs. Generalization

The trend in all fields of agriculture seems to be toward specialization. In agronomy, the specialization is in genetics and plant physiology, rather than the more general areas of crop production and soil fertility. The animal science professionals specialize in genetics and feed nutrition at the expense of the more general area of livestock production and management.

A series of seminars was presented to the animal husbandry faculty at the Sichuan Provincial Agricultural University in Yaan, China (Esmay, 1986). The general topic of the seminars was the utilization of appropriate technology in the management of livestock environment and surrounding conditions for economic returns. After one or two seminars, it became apparent that the faculty had very little interest in management, and even less knowledge about it. This was particularly evident during a guided tour of the livestock facilities at the University. The traditional stall-type dairy barn had just been cleaned by hand and then washed down with an abundant supply of water, thus it was damp and cold and the animals were soon soiled. Further discussions with the faculty revealed that they had no concept of air-moisture relationships with the interactions of temperature and air movement.

The interaction of moisture, temperature, and comfort was even more generally misunderstood as evidenced by the cleaning procedures in the University's lecture halls. The month was November and no heat was being provided, thus the lecture room temperatures were only slightly above freezing. At the end of daily classes, the hallways and lecture rooms were washed down with water. The little heat that was available in the building was utilized to evaporate and eventually dry the water. In the meantime, the building environment was damp and colder than it would have been without the water.

Specialization in Agricultural Engineering. The profession of agricultural engineering, like all engineering fields, is a service profession and should serve the needs of society in the development process. More than 80 years ago, when the American Society of Agricultural Engineers (ASAE) was formed, the main focus was on improving field machinery for crop production. Fortunately, the society name included "agricultural engineering," rather than the more narrow term of "machinery." This allowed for a convenient and necessary broadening of the profession as agriculture and food consumption habits changed to include soil and water, buildings and environment, and rural electrification. More recently, electric power, processing, and food engineering have been added.

Many other countries of the world, such as Japan and China, began by including "agricultural machinery" in their society name. As a result, they have experienced tremendous difficulties in incorporating other agriculturally related engineering fields into one society. In fact, the fields of soil and water and food engineering have ended up in separate societies. The most unfortunate aspect of this situation is that the teaching curricula and research activities also were separated, resulting in more specialization at the undergraduate level.

Most American agricultural engineering departments offer just one basic undergraduate curriculum (possibly mandated by economics and efficiency) in the broad field of agricultural engineering. Some electives are normally offered to strengthen an area of interest, but are not sufficient for appreciable specialization. Specialization, if desired, must be obtained at the postgraduate level. Again, unfortunately, many of the agricultural engineering institutions around the world have not followed this pattern of non-specialization at the undergraduate level. Separate curricula have been developed for power and machinery, soil and water, etc. In some cases, faculties (schools or colleges) of agricultural engineering have been organized with separate departments for each area of specialization. The extreme example of specialization is the Beijing Agricultural Engineering University in China, which has some 17 different curricula within agricultural engineering.

The real-world engineering problems of agricultural production and processing do not fit neatly into one of the traditional specialized areas. For example, the development of a tillage machine is not just a machine design problem. The objective of the machine is to enhance plant growth by helping provide a better micro-environment of air, soil, and water around the plant. Other problems that must be considered and solved pertain to soil-water retention, minimization of soil losses, coping with soil cover materials, soil compaction, draft requirements, economics, and overall appropriateness for small, as well as large farms.

Many private and public sector leaders were interviewed in Zimbabwe (Esmay, 1986) as a part of an agricultural engineering manpower study. The recurring response to the question about the type of training needed was that it should be non-specialized within the broad agricultural engineering field. Respondents emphasized that graduates must have a good working knowledge of the basic biological and engineering sciences, and be qualified to implement effective and innovative design and development of appropriate technology for the agricultural industry of Zimbabwe. Such design and development of appropriate physical technology requires consideration of the many parameters of the fragile ecology of Zimbabwe, including the environment, ecology, types of soils and crops, rainfall patterns, management, and economics, as well as the socio-economic patterns.

Engineers traditionally specialize in analysis and design with a strong focus toward manufacturing and construction. This focus has allowed individuals in other fields to make the early and critical feasibility and planning decisions on the whats, whys, wheres, and even hows for many of the major physical technology type projects in the world. Then, the engineers are called in and told to design the dam, canal, plant or other project structure in accordance with the specifications. In some cases, important physical (engineering) factors have been overlooked or misinterpreted in the decision-making process and made the project infeasible.

Undergraduate agricultural engineering curricula must be directed more generally toward application and utilization than theory and research. This philosophy concerns some who believe that the curriculum might not meet the "engineering criteria" for acceptance as an "engineer," whatever that may be. Others, particularly instructors with specialized training, had rather teach theoretically based courses from the text rather than subject matter that relates to real-world problems in their country. Traditionally, successful engineering graduates, employed either in the private or public sector, move toward management and supervision as their careers progress. Only a broad undergraduate curriculum can prepare the engineer for the transition of these career responsibilities.

Postgraduate Education

Students from developing countries who study in America generally enroll at the postgraduate level. Advanced degree programs, thus, have the greatest impact on the developing country students. The generally quite flexible postgraduate-level programs provide some opportunity for the developing country students to broaden their training if they come from a specialized undergraduate program. More ideally, the MSc provides the opposite, which is the opportunity to specialize somewhat and obtain exposure to research methodology. The MSc degree significantly improves the student's qualifications for many positions in the industry and the public sector. However, this degree is now becoming less considered as a terminal degree, particularly for the education and research fields.

Ph.D. Specialization Vs. a Broader Systems Approach. The doctoral degree in engineering has traditionally consisted of advanced theoretical studies in a narrow area. This was deemed necessary to fulfill the Ph.D. degree requirement for original research in some area at the forefront of knowledge. Such a program leaves little opportunity for any generalization or expansion into such areas as systems science. Doctoral programs that are entirely theoretical do not provide the most advantageous and usable knowledge and educational training for international students from developing countries.

Developing country students do not generally have high technology laboratories available to continue their Ph.D. dissertation research when they return to their home country. Even if the facilities were available, the research would undoubtedly not be the most appropriate for that area of the world. Most developing countries need more applied research to find solutions to critical real-world problems in food and fiber production and processing. The teaching requirements for the Ph.D. graduates returning to their home country are somewhat similar because the undergraduate courses they instruct should not be highly theoretical or narrowly focused.

Ph.D. graduates, after returning home, also move very rapidly into administrative and supervisory positions (Esmay, 1989). They are talented, highly qualified individuals in a country with enormous shortages of trained professionals, so rapid advancement is normal. Theoretical and specialized training at the doctoral level do not adequately prepare these individuals for the duties of administration and management. A liberal mixture of system science, management, and economics courses along with their engineering curricula, would better prepare these professionals for the array of responsibilities with which they are confronted. The systems approach provides individuals with the capability of considering and analyzing all relevant parameters in problem-solving.

Non-Engineering Degrees. It is a paradox that the classic engineering degree does not provide the most appropriate educational training, particularly for individuals from developing countries. Most American agricultural engineering departments offer non-engineering degree programs at the BSc and MSc levels. They have been called by different names—from agricultural mechanization to agricultural engineering technology. These programs have greater breadth, with the emphasis on application and utilization. They unfortunately have not been popular programs in most cases because they have gained the reputation of being for students who cannot succeed in more rigorous engineering programs. The students and graduates of the non-engineering programs often are concerned that they will not have the prestigious engineering title. Nevertheless, graduates have excelled in the agro-industry fields.

The Michigan State University Agricultural Engineering Department expanded the non-engineering program to include the Ph.D. in systems science, and computer modeling provided the necessary rigor for research at the doctoral level, as well as the desired breadth for the holistic approach to problem-solving. Although a number of international students have undertaken the program, many hesitated because the program did not provide the engineering title. In some cases, the choice for the engineering degree was the student's choice. Essentially, all developing country departments believe that they must graduate "agricultural engineers" rather than non-engineers referred to by such titles as agricultural mechanization or technology. Even so, non-engineering graduates can readily handle 90 percent of the physical technology problems of their developing country. Also, in many cases, the departments are ill-prepared in faculty and resources to offer a bonafide engineering curriculum. In this situation, graduates are called engineers but really don't have the equivalent training, and cannot be admitted into graduate agricultural engineering courses abroad. In addition, they are poorly prepared to perform engineering design and analysis when hired into an engineering position.

In-Country Research for International Students. The systems approach is necessary for doctoral graduate students to conduct their dissertation field research outside of the U.S., as it provides flexibility for monitoring, modeling, and analyzing physical agricultural production and processing systems in developing countries, regardless of the level of technology. The opportunity to work on dissertation research topics relevant to their home country is of great value to developing country students. A number of the doctoral graduates with this in-country research experience have continued to conduct related research and development upon completion of their studies and return home because it was relevant to the needs of the country.

Why isn't in-country dissertation research implemented by more international graduate students? Probably because a significant amount of time and effort is required from the major professor and graduate student to make all of the arrangements. In addition, international travel and research adds costs. International field research requires 6 months to a year and, in most cases, extends the overall time for the doctoral program as compared to an on-campus, U.S. domestic project. Following the in-country field research, the students return to the campus to conduct the analysis, modeling, and preparation of the dissertation and final defense. These activities normally require another 6 months to a year. However, results have been excellent, and the return on investment is worth the extra time, effort, and money. Such programs are recommended and should be pursued whenever and wherever possible.

General Vs. Specialized. The agriculture of the future, with emphasis on sustainability, requires broadly trained professionals in all agricultural fields. The systems approach and modeling is applicable to all of the various disciplines of agriculture to make problem-solving possible on a holistic basis. Agricultural economists were the first to enter the systems and modeling fields and have been successful in the application and incorporation of these analytical tools into their agricultural discipline.

Future International Development Strategies

Smuckler et al. (1988), in a report resulting from a broad year-long study of many agencies and top internationalists, states that the following three cardinal objectives should be followed: (1) broadly based growth; (2) an effective attack on poverty; and (3) an end to the destruction of the environment. They state that "To accomplish these objectives, the United States must forge a new, a more mature relationship with the Third World. We must shift from the old idea of aid, to the new idea of mutual gain through cooperation. With developing countries as partners, the United States can progress together into the 21st Century."

"Why is now a good time to make major changes in our engagement with the Third World? Because the world has changed in the past several years—in global economics and politics, in our domestic economics, as well as in the Third World itself. The Third World is no longer a single homogeneous group of countries. Now they range from the very poor to the newly industrialized " (Smuckler et al., 1988).

The United States should use its experience and skills to accomplish the following:

- To enhance physical well-being through improved health systems and population planning.
- To work for sustainable agricultural systems, particularly emphasizing food supplies and forestry.
- To develop environmental programs and policies that will protect natural resources, and, by emphasizing renewable supplies and conservation, ensure better energy security.
- To foster sound urban development policies.

To carry out these challenges, the Smuckler report, emphasizes that "The Agency for International Development must change. Its structure and name ought to reflect the new theme and style of mutual gain through cooperation. To this end it should expand its analytical capacity and guide U.S. agencies toward a new U.S. relationship with the developing countries."

The Smuckler report challenges the new administration with the following:

- Establish a council led by the White House to coordinate U.S. agencies' actions and policies on development.
- Create a semi-autonomous foundation to strengthen research and the use of science and technology for development.
- Consider the size and number of overseas missions as new tasks and cooperative modes evolve.
- Involve intermediaries more often and use binational councils and boards in countries abroad.

In summary, the Smuckler report emphasizes that some critical changes must be made by the U.S. in its future relationship with the developing Third World countries. The approach must be on a mutual cooperative basis rather than the past philosophy of strictly providing aid. Science and research is emphasized as an important input to future programs with the Third World. Maintenance of the environment and natural resources has displaced food production as the first priority; however, food security is still vital in food-scarce countries such as Africa.

Summary

A worldwide movement toward sustainable agriculture is necessary, with top priority given to the prevention of environmental degradation and the waste of natural resources. This requires a shift from the short-term past and still somewhat prevalent present emphasis on maximizing crop production (yields), to longer-term, broad consideration of all related environmental factors. Crop production inputs must be drastically reduced, particularly environmentally detrimental chemicals and non-replaceable energy (petroleum products). Minimizing soil losses must be a top priority consideration for all cropping systems.

The systems approach must be integrated into all agricultural teaching and research programs. Today's complex and interrelated problems can only be satisfactorily solved through broad approaches that are integrated and consider all parameters. Teaching and research programs must be directed toward solving the real-world problems of local, in-country environmental and food production systems. A more general education, along with the systems approach, will better prepare professionals for broad, integrated approaches to problem-solving.

8

The Cultural Ecologist Concept of Justice

Billie R. DeWalt

Introduction

Much contemporary debate centers on how effective the agricultural research system has been in developing new technology for agriculture and helping humans to better adapt to social and natural habitats. Phrased in a contrasting way, the issue is whether or not: (a) agricultural research has enhanced productivity so the life-giving forces of natural resources and the welfare of human inhabitants of the planet have been improved; or (b) agricultural research has served special interests pursuing profit, thereby leading to environmental degradation and overlooking the needs of a large portion of the population.

The purpose of this chapter is to examine critiques of the international agricultural research system. I believe that, from the inception of efforts to apply science and technology to solve agricultural problems in the developing world, important social, moral, and ethical questions have been raised. I also argue that these issues generally have been ignored or "explained away" by those leading such efforts. I will suggest some reasons why these issues have been disregarded and conclude by recommending some new positive steps to more effectively incorporate moral, ethical, and social justice issues into international agricultural research.

Internationalizing Agricultural Research: Early Stages

The real beginnings of the current international agricultural research system[1] can be traced to the joint program on food crops initiated in

1943 between the Rockefeller Foundation and the Mexican government (Baum, 1986). The purpose of Rockefeller's Mexican Agricultural Program (MAP) was: "...to increase the production of varieties, the improvement of the soil and the control of insect pests and plant diseases." A corollary goal was to "train young men and women in agricultural research and in the development of techniques for promoting the rapid adoption of the new technology " (Wellhausen, 1976).

The program was first staffed primarily by Americans, but as the program continued, more and more posts were filled by Mexicans returning from graduate studies in the U.S. At the same time, researchers from other Latin American countries also were being trained in the U.S., and the U.S. began assisting in the improvement of national agricultural research programs and in the establishment of regional organizations to promote interaction (DeWalt & Hudgens, 1988).

The MAP and its successors, the Mexican National Forest, Agricultural and Livestock Research Institute (INIFAP), and International Maize and Wheat Improvement Center (CIMMYT) are important to the history of agricultural research because their efforts originated the first attempts to apply research breakthroughs in U.S. and other Western agricultural areas to developing parts of the world. This effort established the precedent for creating the system of international agricultural research centers (Plucknett & Smith, 1982; Baum, 1986), now numbering 13. The MAP and the International Rice Research Institute (IRRI) in the Philippines also developed the so-called "miracle seeds" that created what has become known as the "Green Revolution."

Many of the later critiques of this international agricultural research system were forecasted during the beginnings of the system. Several of the criticisms were quite prophetic and merit some discussion.

Even before the program began, an outstanding cultural geographer from Latin America, Carl Sauer, was asked by the Rockefeller Foundation to make recommendations for assisting Mexican agriculture. In 1941, he cautioned against the attempt to recreate the model of U.S. commercial agriculture in Mexico:

> A good aggressive bunch of American agronomists and plant breeders could ruin the native resources for good and all by pushing their American commercial stocks...Mexican agriculture cannot be pointed toward standardization on a few commercial types without upsetting native economy and culture hopelessly. The example of Iowa is about the most dangerous of all for Mexico. Unless the Americans understand that, they'd better keep out of this country entirely. This must be approached from an appreciation of native economies as being basically sound (Jennings, 1988).

Sauer also emphasized that the nutritional and agricultural practices of small Mexican farmers were quite sound and that their main problems were economic rather than cultural. In 1945, Sauer followed up his earlier warnings by noting that the MAP program research on maize could be beneficial, but that the research on barley, wheat, and alfalfa probably had limited or no benefits for local communities. Furthermore, he warned that the tendency in Mexican agriculture was beginning to mirror that of California: "...interest is directed away from subsistence or village agriculture to the needs of city and factory with the attendant emphasis on standardization of product and on yield...and on the commodities which the privileged fraction of the population can afford " (Jennings, 1988).

The warnings of Sauer were dismissed by influential people in the Rockefeller Foundation as the concerns of someone who thought of Mexico as "as kind of glorified ant hill," was concerned about the destruction of "picturesque folk ways," and thus resented "any effort to 'improve' the ants" (Jennings, 1988).

The second related problem that was addressed concerned the political, economic, and social effects of the new technologies being developed. A report was prepared in 1949 by a group led by John Dickey, then president of Dartmouth College, who noted the following:

> For example, I can imagine that this program before long might begin to have a considerable impact upon the whole land-use policies of Mexico, and I am perfectly sure that within three to five years the program will raise some very acute problems with respect to the political control of these benefits...these very benefits may introduce fresh economic disparities within the Mexican economy, which will present political problems not now even dimly perceived by many Mexicans (Jennings, 1988).

Rather than confronting these problems with research and other measures, Dickey's recommendation was to avoid the issue. He wrote: "It would be unfortunate for all concerned, especially for the program itself, if the Foundation is heavily in the picture when this [growth in social tensions] takes place " (Jennings, 1988).

In summary, critiques from the 1940s focused on several issues that continue to be important in debates about the effectiveness of the international agricultural research system:

- Questions were raised concerning whether technology creation was based on the wrong premise. Sauer believed that the major obstacle for the social progress of Mexican peasants was economical, not cultural or technological.

- The second issue involved determining for whom the technologies were being created. Were they designed to meet the needs of the privileged fraction of the population rather than the economic and nutritional needs of the poor?
- What would be the social, economic, and political results of the technologies being created and diffused?

More recently, the international agricultural research system has begun paying more attention to these issues, rather than ignoring them as it did in the 1940s and 1950s. The following section assesses these recent trends in international agricultural research, and discusses difficulties, why difficulties remain, and how they might be resolved.

More Recent Trends in International Agriculture

Until the 1960s, most social scientists shared the same goals and assumptions of biological agricultural scientists. Most believed that "modernization" and technology transfer were necessary for the Third World countries to experience a "take-off" into sustained growth (Rostow, 1963; 1971). Social scientists concentrated their efforts on such research agendas as the diffusion of innovations (Rogers, 1969; 1971). This involved examining how and why technology was adopted, identifying the cultural obstacles to change, and determining why people continued to cling to traditional behaviors and technologies (Banfield, 1958; Foster, 1967; Lerner, 1958; McClelland, 1961).

Critiques of U.S. agriculture began to appear[2] with Rachel Carson's *Silent Spring* (1962), Barry Commoner's *The Closing Circle* (1971), *Hard Tomatoes, Hard Times* by Jim Hightower (1973), and *The Unsettling of America* by Wendell Berry (1977). These were paralleled in the international arena by the dependency theory perspective advocated by Andre Gunder Frank (1969), Samir Amin (1976), and others, and the critiques of the Green Revolution by people like Keith Griffin (1974), among others. The overall conclusion of these critiques regarding agriculture was that much of the new technology being created led to greater benefits for larger farmers, but did not increase welfare in rural areas. The benefits of new agricultural technology provided cheap food for growing urban areas of the Third World.

In conjunction with the criticisms of the Green Revolution were those who identified the replacement of semi-subsistence crops by cash crops as a significant problem in the developing world (Lappe & Collins, 1977; Lappe, Collins, & Kinley, 1980; Fleuret & Fleuret, 1980). These individuals pointed out that small farmers who adopted cash cropping were drawn into exploitative relationships in the market system. They often ended up worse off, nutritionally and economically, for adopting cash crops. At the micro level, their arguments reflected the philosophies of dependency theorists who believed that the terms of trade consistently worked against agricultural products from developing countries. The consequences were that the poor continued to be exploited by the rich, and rural citizens abandoned their homes and migrated to the cities.

Defenders of the Green Revolution have continually contested the claims made against it, and only in the late 1980s, have more balanced appraisals surfaced (Lipton & Longhurst, 1989). While they are not as negative about the Green Revolution, Lipton and Longhurst conclude that technical breakthroughs alone cannot solve the problems of social and economic inequality. Similarly, balanced assessments of the debate over cash vs. subsistence crops are now appearing (von Braun & Kennedy, 1988; Maxwell & Fernando, 1989; DeWalt et al., in production).

In general, however, many are recognizing that the process of development as it has occurred during the past four decades has problems. Despite the affluence that exists in the U.S., development has not resulted in a significantly better life for millions of people. In the international arena, despite three decades of development proclaimed by the United Nations and a long history of multilateral and bilateral assistance, the World Health Organization estimates that more than one of every five humans—about a billion people—is suffering from disease, poor health, or malnutrition. Furthermore, the development of both industrialized and lesser developed countries has often resulted in water, air, and soil damage; massive deforestation; desertification; and ecological disasters of all kinds. The inequalities between rich and poor, both within and across countries, continue to grow larger.

Thus, the two major criticisms about development approaches until now may be summarized as follows. First, development has frequently been accompanied by severe negative environmental effects—what I call the relationship between culture and nature. Development often leads to destruction of large ecological systems (Bennett, 1976; DeWalt, 1988b). Secondly, development has frequently been accompanied by negative effects on human relations. For example, development has often led to the destruction of cultural systems (Bodley, 1990). Cultural diversity has been increasingly destroyed and the world's cultures have been homogenized (DeWalt, 1988b). In addition, pervasive inequalities continue in the world. Most glaring is the paradox that worldwide agriculture is plagued by surpluses at the same time that a fifth of the world's population is malnourished.

Why haven't the issues raised by the critics of agricultural development been given more credence? Why haven't the concerns been incorporated earlier? Why are we so incapable of pulling moral, ethical, and value questions into the process of development?

The Problem

A significant part of the problem relates to several basic elements of the scientific belief system. These beliefs, combined with other institutional and social policies that have ignored the environmental and human relations aspects of development, have created many of the existing problems.

The first aspect of this belief system is the faith placed in technological determinism, or the technological fix. Throughout this century, the pervasive belief has been that technology is the ultimate answer to world food and poverty problems. This faith in the efficacy of technological determinism is reflected in the way we refer to the "revolutions" created by the advance of technology.[3] The "Agricultural Revolution," "Industrial Revolution," and "Green Revolution" are all considered hallmarks of human progress, made possible by the advancement of technology. Although none of these earlier revolutions have solved poverty and starvation problems, many individuals now believe that the emerging "Biotechnology Revolution" is the answer to world food problems. By producing more food more efficiently, the perception is that enough will be available for everyone.

We must recognize, however, that the earlier revolutions have been extremely disruptive in their effects on society. Jared Diamond, for example, has stated that the adoption of agriculture may have been "the worst mistake in the history of the human race." He reports that: "Forced to choose between limiting population or trying to increase food production, we chose the latter and ended up with starvation, warfare, and tyranny" (1987). Instead of producing a better life, the Agricultural Revolution was "...a catastrophe from which we have never recovered. With agriculture came the gross social and sexual inequality, the disease and despotism, that curse our existence" (Diamond, 1987). Diamond is also a technological determinist, but rather than believing that technology has affected the human race positively, he sees the agricultural revolution as putting us on the path toward the misery that currently exists.

Viewed in these terms, the Industrial Revolution, which led to extremely high rates of economic development in Europe, for example, was also accompanied by significant social problems. During the 8 decades from the middle of the 19th century to the first quarter of the 20th century, some 48 million people (about 12 percent of the population in 1900) left the continent (Massey, 1988). Infant abandonment by families who could not afford to take care of children was so common that foundling homes were common in every European country.

In the same way, the mechanization of agriculture in the U.S. led to the mass migration of millions of people to urban areas. We tend not to focus on the misery and suffering caused by these processes, perhaps because we and our ancestors have been survivors rather than victims of them. In general, historians have not recounted the stories of individuals who suffered the effects of these earlier revolutions.

Will the Biotechnology Revolution avoid some of the dislocations caused by the earlier revolutions? The potential exists to achieve far greater production with fewer people. But a billion people in the world already are not adequately clothed, fed, educated, and healthy because, as our economic systems are currently organized, there is nothing productive for them to do. Are dislocations and disruptions necessary adjuncts of technology development and diffusion? I think not, but if we want to avoid them, we must recognize that *the way technology is created and implemented must have values, morals, ethics, and social justice issues incorporated from the beginning.* Technology makes many things possible, but it is the social, political, and cultural context that determines what effects technology will have on the world.

A second reason why these aspects have not been incorporated earlier involves the implicit assumption that science (in this case, as exemplified in agricultural research) is value-free and socially and culturally neutral. In social science and the humanities today, all kinds of intellectual trends (deconstructionism, relativism, hermeneutics, postmodernism, etc.) have as their fundamental premise that all knowledge is constructed through the interplay between the observer and the observed. The important message to derive from these perspectives is that any research (in any discipline) is fundamentally part of the political process. The grant and contract process, our own education, and our participation in a culture and a nation, all help determine our research problems. Inevitably, we serve some masters and not others. As scientists, we need to engage in a frank self-evaluation of our efforts and recognize who we are serving.

This leads to a third reason why values, morals, ethics, and social justice issues have not been considered in the development process. The politics and structure of biological agricultural research have actually created some of the problems that we see in international agriculture today. One aspect of the structure that is extraordinarily important is the problem of specialization, both within and across disciplines. Most scientists today work on basic research issues that address very specialized topics within fairly specialized disciplines. This means that scientists are far removed from the ultimate outcome of their research; they are like factory workers who put one part into the whole but who may never recognize what the end product is.[4]

In addition, scientists are encouraged and rewarded for working on topics that are on the "frontiers of knowledge." That is, researchers are not rewarded for tinkering with existing technology or plants to make them better or more productive. Unfortunately, research structures place greater priority on fashionable topics than the problems of the poor. A bias toward newness and technological complexity is prevalent. Researchers are not given the same kind of academic rewards or other recognition for applied research as for basic research. One of the trends that Busch and Lacy (1983) have documented for every agricultural discipline is that throughout time, each becomes more like a basic "scientific" discipline with journals and academic meetings that would be incomprehensible to any lay person.

This specialization, especially of language and topics deemed to be important in the discipline, creates separate "cultures" and makes interdisciplinary and multidisciplinary work difficult, if not impossible. While working as part of the International Sorghum and Millet Collaborative Research Support Program (INTSORMIL), for example, I was surprised to find almost as much hostility among the agricultural scientists as between them and the social scientists (DeWalt & DeWalt, 1989).

A final aspect of this specialization problem is that those who create the technology are not the marketers of it. Commercial firms or establishments must take basic research and turn it into a final product. Their concern is obviously to make a profit with the technology; social, moral, ethical, and social justices are not their responsibility. As the Mexico case discussed earlier in the chapter illustrates, this same attitude has, to some extent, been adopted by the international agricultural research centers. In part, to shield themselves from criticism, they explicitly state that they are involved in the production of intermediate goods—germplasm, training, and other expertise—and that national programs are responsible for the production and dissemination of the final technologies. Few national programs have this expertise, which means that commercial firms are provided with an opportunity to step into the void. And these firms are unlikely to put social justice or environmental concerns high on their agendas.

Toward Equitable and Sustainable Solutions

What should be done? Scientists must recognize that they are a part of the problem as well as the solution. Agricultural scientists are certainly correct when they say that governmental and institutional policies that provide the context in which the results of agricultural research are applied are often to blame for the negative environmental and anti-poor effects of technology. At the same time, however, the kinds of technologies that would allow for progressive moral, ethical, and social justice dimensions are in too little supply in the international arena. For many of the reasons listed previously, the agricultural research needed to address these concerns is considered to be outside the realm of "normal science."

A major aspect of the recognition problem is for scientists and policy-makers to get away from what we may call the "accidental tourist" phenomenon. These individuals often take field trips to communities frequently during the best season of the year, and visit the fields of a few model farmers (usually near a road), take a few pictures, and return to their laboratories, experiment stations, or offices with an "understanding" of the needs of farmers. Agricultural scientists and those who develop and implement public policy affecting farmers must have a much greater awareness of the reality of the small agriculturalist. If they are unwilling to obtain this experience firsthand, then they must read the publications of social scientists who have worked in the region of concern.

The emphasis on farming systems research and extension that was predominant several years ago had great promise as a perspective that would encourage multidisciplinary projects among social and biological agricultural scientists. While its influence is predictably waning (DeWalt, 1985), a victim of the changing fads of development practices, significant emphasis is being placed on sustainability. While this emphasis may eventually prove to be as unsustainable as farming systems research, sustainability may be interpreted in ways that will incorporate some of the moral, ethical, value, and social justice dimensions that are critical for more equitable development.

However, some aspects of the sustainability emphasis are somewhat disturbing. Current definitions and discussions of sustainability focus principally on environmental stewardship—human use that minimizes pollution, depletion, and disruption of the ecological balance of natural resource systems. In other words, emphasis is on meeting the immediate needs of human communities without jeopardizing the ability to meet future needs. This definition focuses on the environment, conveniently omitting the needs of human populations. Sustainability easily can become a way for scientists and researchers to address the associations of physical resources without confronting the political, economic, social, and cultural dimensions of development.

Sustainable development should involve at least the following four dimensions:

- **Environmental stewardship:** As defined previously, human use that minimizes the pollution, depletion, and disruption of the ecological balance of natural resource systems; meeting the immediate needs of human communities without jeopardizing the ability to meet future needs.
- **Community:** Methods that minimize conflict in the use and appropriation of natural and manufactured resources. Central in this dimension is the issue of economic and gender equity because sufficient equity in the access to and distribution of resources is necessary to ensure political stability.
- **Diversity and participation:** The means by which cultural and ecological diversity can be fostered and maintained (DeWalt, 1988b), and appropriate ways to empower and foster participation of affected groups to determine how resources are appropriated.
- **Financial and institutional:** Allocation of financial, human, and institutional resources to increase the long-term wealth and well-being of affected populations; determining how private enterprise and the public good can best be reconciled.

These goals, which are consistent with the "ten precepts for survival" that I developed several years ago (DeWalt, 1988b), recognize that people are part of the ecology too. Sustainability is not possible until poverty is eliminated because poverty is one of the primary causes of overpopulation, environmental destruction, crime, malnutrition, and poor health.

The structures are in place to allow and encourage the development of agricultural technology with a human dimension. One important element can be the land-grant universities and the Title XII program that is attempting to make them resources for international development. But a different set of priorities must be established. The client group should be defined as the poor, and the goals of research should be poverty alleviation and development of methods to improve fulfillment of basic human needs and ecological stewardship.

Unfortunately, biological agricultural scientists emphasize the pressures they feel to serve the interests of commodity groups and farmers in their own state. Yet Gillespie and Buttel (1989) recently have shown that most farmers in New York do not believe the research from Cornell benefits them. In fact, they are quite suspicious of the research efforts. Some may argue that the U.S. land-grant agricultural research system may have eliminated its clientele by developing technology to shrink the farm population to about 2 percent of the country's total.

Conducting research on international agricultural issues will undoubtedly cause some difficulties for colleges of agriculture because remaining U.S. farmers may claim that we are aiding the "competition." The structure of the U.S. university system, however, is designed to protect the rights of scientists to pursue the areas of research they believe are most needed. The tenure system should allow agricultural scientists to do the right thing and work on the truly difficult problems that face the resource-poor farmers of developing countries.[5]

We should strive for some balance in the U.S. agricultural research system. Money from private corporations is available to some researchers who continue to develop technology for larger, resource-rich farmers. Public funds should be allocated to work on the problems of smaller, resource-poor farmers, both in the U.S. and the Third World. Biotechnology has the potential to create plants for resource-poor areas of the world. Yet one can argue that the majority of the research now under way benefits large farmers. That task should be left to large corporations. We need a greater division of labor and responsibility; the wealthy farmers will have technology developed for them, but the more pressing and difficult problems exist among the world's resource-poor farmers.[6] Scientists supported by the public sector and multilateral and bilateral aid agencies must recognize that their responsibility is to put moral, ethical, and social justice issues at the top of their agendas.

One method by which U.S. universities can be reorganized to better meet these needs is to create departments, institutes, or centers with the primary responsibility of international agriculture or international development. What is needed is an institutional arrangement in which faculty with international development interests can achieve the following:

- Conduct research that will be rewarded because of its contribution to the solution of international problems (rather than in spite of it, as is now so often the case).
- Interact in a multidisciplinary setting with other faculty who have similar goals and interests.
- Train future generations of scholars and researchers who are even better equipped to address the difficult problems of development. These institutional structures should include both biological and social scientists, and educational programs should require biological scientists to complete several basic courses in humanities and social sciences, while social scientists attend basic courses in biological sciences. The core of such a program should be a seminar that addresses the ethical, moral, value, and social justice dimensions of development.

Internationally, the International Agricultural Research Centers (IARCs) are an appropriate location for research targeted to the needs of the poorer regions of the world. Yet the research in the IARCs now often duplicates the specialization and "frontiers of knowledge" research underway in U.S. universities. Many of the attitudes and perspectives about the social, political, and cultural effects of technology generation that were outlined at the beginning of this chapter still remain in the IARC system.

To make progress in improving the world's food and nutrition situation, interactive efforts between biological and social scientists in U.S. colleges of agriculture and in the IARC system are critically important. What must be recognized is that both groups of scientists are on the same side and working toward the same goals. In the past, much mistrust has existed between these groups. Some social scientists have made their careers on criticizing the work of biological agricultural scientists. As a result, social scientists have been excluded from working in agricultural development institutions. However, social scientists must receive greater responsibility for planning, monitoring, and evaluating agricultural development projects. Social scientists should identify problems and, in conjunction with humanists, examine the ethical, moral, and social justice dimensions of the potential solutions. Biological agricultural scientists can then design technical solutions more knowledgeably.

Notes

1. The origins of international agricultural research can be traced to the age of exploration and discovery in the 15th and 16th centuries. The contacts established between the Old and the New World led to what has become known as the Columbian Exchange (Crosby, 1972), the widespread interchange of plant and animal species. World-famous Kew Gardens in England began collecting many species of tropical plants, partially to determine their potential economic importance. Early agricultural research stations in the conquered countries were established for explicitly commercial, colonial purposes—to assist in introducing and producing agricultural products that could be grown for the benefit of entrepreneurs from the colonizing countries.

2. Walter Goldschmidt's work (1978; orig. 1947) on California agriculture, *As You Sow*, had been completed in the late 1940s, but it was suppressed and did not attaract much attention until the "age of protest" began.

3. Jack Kloppenburg has cited some other positive claims for the fruits of agricultural research (1988). For example, hybrid corn varieties have been credited in congressional testimony for paying for the development of the atomic bomb (Shull, 1946) and Paul Mangelsdorf credited hybrid corn with containing the spread of Communism because it provided enough food for Western Europe after World War II (Glass, 1955).

4. Even plant breeders who create and recreate whole plants are somewhat specialized in these terms because they may not be cognizant of the cropping system into which their particular species fits. Unfortunately, one of the consequences of this situation has been the trend toward monoculture, in which any "ideal" environment is provided to gain maximum productivity from one species.

5. Similar problems exist in the medical arena. Malaria, diarrhea, acute respiratory infections, and other diseases cause many millions of deaths each year, yet the amount of research on them is minimal compared to work on affluent diseases such as heart problems, cancer, and others. Two few medical researchers are focusing on the problems that afflict the poor. The poor are unable to pay, thus private drug companies have little interest in their problems.

6. A good example of attempting to achieve balance is the current research on sorghum in Latin America. Sorghum is now being grown as a livestock feed crop in many countries of Latin America (Paul & DeWalt, 1985; Barkin et al., 1990). Pioneer, DeKalb, Northrup, and many other private companies have moved into this market with improved hybrids, and farmers' yields have been quite good. At the same time, in Haiti and Central American countries, sorghum is an important food crop for poor people and yields are desperately low (DeWalt, 1985). Projects in the International Crops Research Institute for the Semi-Arid Tropics, the International Sorghum/Millet Collaborative Research Support Program, and national programs in several countries focus primarily on developing technology for food-quality sorghum produced by small farmers.

PART TWO

Bibliographic Essays on the
Critics of Agriculture

9

Agricultural Research Policy and the Family Farm

Nancy W. Shankle

Introduction

Agricultural research policy is concerned with how scientists choose their research problems; how projects are approved; how funds are distributed; how research is disseminated through extension agents, patents, and journals; and generally how to define the benefits of research. Agricultural research decisions often have specific effects on the family farm. The purpose of this chapter is to provide a guide to significant sources of information relating to agricultural research policy and the family farm. The review of literature is selective because of space limitations and the inefficiency of an unrestricted discussion of the material. This text is intended to serve as an introduction to agricultural research policy and the family farm for the uninformed reader who may have a reason to investigate the subject. It is not intended to serve as a substitute for an examination of the literature itself.

Scope

Few of the articles found in the review of literature directly link agricultural research policy with the family farm. A number of articles discuss agricultural research policy in detail, giving a few implications about the family farm, while others discuss the family farm at length, concluding with a few comments on new directions for agricultural research. Still others discuss government farm policy and relate to agricultural research policy only by suggesting that research policies be consistent with government policies.

Given this variation in literature, sources which did not directly address research policy and the family farm have been carefully selected. Some articles were chosen because they were seminal pieces about a given issue; others were chosen because of the ability of the author(s) to capture the depth and breadth of the discussion with fluency. Full citations for all pieces discussed in the text are included in the bibliography at the end of the book; additional references, not discussed in the text but nonetheless relevant, are also cited.

The literature review is organized into three major sections and two representative case studies. First, are definitions and histories: The difficulty of defining the family farm, the variations of agrarian themes in America, and the historical patterns of family farms and farming ideals are explored. Arguments for or against preserving the family farm are organized around economic, environmental, and philosophical themes. The second major section deals with the development of agricultural research policy. Again, the discussion begins with the history of agriculture in the 19th century, particularly with the establishment of land-grant universities, experiment stations, and the extension service system. A discussion of present-day research policy is provided, including how policy is formed, and criticisms of current policies. The third section is a discussion of one important outcome of research, technology, and the role it has played in maintaining or undermining family farms.

Definition and History

The idea of the family farm began in American history in the 18th century and is usually traced to Thomas Jefferson. Richard Kirkendall in "A History of the Family Farm" (1987a) argues that Jeffersonian agrarianism was a blend of aristocratic agrarian ideas and democratic politics. By the 20th century, two clear movements had developed: the populist movement, which sought to reassert agrarian values; and the modernization movement, represented by the USDA, land-grant universities, railroads, etc. These movements both emphasized the importance of agriculture, but nevertheless represented distinct philosophical differences. The agrarian movement proclaimed the political and social importance of farms and farmers, while the modernization movement stressed the economic significance of farming and formed systems now called agribusinesses.

Kirkendall further explores the agrarian myth in two other articles. In "The Central Theme of American Agricultural History" (1984), he outlines three major components of the American agrarian theme.

1. The political, which supports the belief that democracy depended on the existence of a large number of independent family farmers.
2. The social, in which the rural environment was considered morally superior to the urban life and necessary to preserve the quality of the nation.
3. The economic, which maintained that the wealth of the nation depended upon a healthy agriculture.

Kirkendall offers an historical perspective to farm problems in his comprehensive essay, "Up to Now: A History of American Agriculture from Jefferson to Revolution to Crisis" (1987b). Here, he concludes by projecting that for land-grant universities to be successful agencies of change, they must broaden their outlook to include all rural interests, not only production and farm business, but also rural society. That is, "the universities must have strength in both the basic and the applied sciences and in the social sciences and humanities" (p. 23). Kirkendall includes an extensive bibliography in this essay.

Paul Thompson, in "The Philosophical Rationale for U.S. Agricultural Policy" (1988d), provides the philosophical background to the agrarian mystique. While agrarian beliefs differ from one person to another, in general, "agrarians believe farmers and farming have special characteristics that are unique to agriculture" (p. 35). The Jeffersonian ideal, often portrayed today as a celebration of the family farm, was actually Jefferson's response to a political situation. At a time in American history when the proposed concept of democracy was to be restricted to wealthy landowners, Jefferson responded by arguing that the availability of land ownership by all Americans would provide a secure basis for a responsible democratic citizenship.

The vision of farming as a way of life, however, is discussed in the more recent writings of Wendell Berry and Jim Hightower. Thompson writes that Berry's arguments, rooted in the Protestant work ethic, celebrate nature and self-reliance typical of 19th century authors such as Ralph Waldo Emerson. Hightower's views of agrarianism, according to Thompson, hinge on his belief that such a democracy must maintain entrepreneurial opportunities for all citizens, not merely the wealthy or well-capitalized. In this manner, Hightower's arguments are reminiscent of populist movements at the turn-of-the-century. Thompson concludes that agricultural policy should be based on a concept of the common good, rather than special interests, without sacrificing individual interests for the common good.

Two other articles which address the philosophical backgrounds of the agrarian myth deserve mention. James A. Montmarquet (1985; 1989) discusses the philosophical traditions of Aristotle, Locke, and

Jefferson, which are used to support the claims of agrarianism. Mont-marquet argues that the claims vary from the weakest—agriculture is an honorable way of life; to the moderate—agriculture is a superior way of life; to the radical—everyone should have access to productive land. In "The American Wisdom Literature of Farming" (1984), Wojcik suggests that Varro, Cato, Crevecoeur, Thoreau, and Berry present a comparison of wise farming and aggressive farming, and thus promote the mythic themes of agrarianism.

In a discussion of the changing relationship between agriculture and society, Joseph Molnar and Lionel Beaulieu, in "Societal Implications of Changes in the Organization of Agricultural Production" (1984), argue that a decline in the number of farmers has resulted in a new minority; yet these farmers, because of the power of the agrarian myth, continue to wield political power in excess of their numbers. The authors predict that the balance of power will continue to shift from food producers to food processors and marketing companies. These shifts in power also have important implications on directions for agricultural research policy, especially in addressing concerns such as resource management.

Francis Moore Lappe also considers the power of the agrarian myth in "The Family Farm: Caught in the Contradictions of American Values" (1985). She notes the conflict between the fact that most Americans believe family farming is a superior way of life and yet remain passive as it gradually dies out. She argues that the reason for this conflict is because Americans consider free market values as superior even to the agrarian myth.

Several sources discuss the rise of agriculture in the 19th century. In his introduction to *The Agricultural Social Gospel in America: "The Gospel of the Farm" by Jenkin Lloyd Jones* (1986), Thomas Graham describes Jones as an extremely popular preacher and lecturer in an age that consid-ered sermons and lectures as a kind of entertainment. The manuscript, left by Jones at his death in 1918, is actually a series of sermons preached and published separately between 1878 and 1917. Together they present a view of agrarian life, focusing on activities such as plowing, sowing, cultivating, and harvesting crops, as well as developing machinery and techniques for performing these tasks. Jones' message was a liberal one, arguing for the moral view that technology is an opportunity for the improvement of society. He, like many during his time, looked to science and technology to ennoble humanity. This collection of sermons illustrates the excitement and hope of the age and is a companion piece to the historical articles of the period.

The rise of American agriculture, is linked to the rise of capitalism in S. Buik Mohammadi's "American Capitalism and Agricultural Development" (1981). Mohammadi writes that colonial America was restricted in its industrial development because of laws established by Britain designed to make America a producer of raw materials and a consumer of finished products. In fact, even though America achieved political independence following the Revolutionary War, it did not achieve economic independence until the foundation of a sound domestic economy in the 19th century. Early 19th-century agriculture was characterized by self-sufficient food farming in the North and cotton production in the South. By the Civil War, agriculture had established patterns of land-extensive mechanized farming methods to make farming commercially successful.

Following the Civil War, four important laws resulted in significant changes in American agriculture: (1) the act that established the USDA; (2) the Morrill Act, which established land-grant colleges; (3) the Homestead Act, which provided 160 acres of public land; and (4) the Pacific Railway Act, which encouraged development of the railway system to transport farm products to urban centers. Mohammadi argues that the public policy represented in these laws encouraged capital-intensive farming methods that were most appropriate for the wealthier, capitalized farmers. The best lands available through the Homestead Act were owned by a few wealthy landowners, leaving marginal lands for poorly capitalized farmers. In addition, the railroads, through monopolistic activities, effectively doubled production costs and lowered profit margins for farms. Government encouragement to increase capital began the cycle of producing more per acre through improved agricultural methods, but earning less per acre because of product surpluses.

Many authors have considered the problem of defining the family farm as a prerequisite for an adequate account of the relationship between the family farm and agricultural research policy. In "Structural Trends in Agriculture and Preservation of Family Farms" (1986), John Lemons argues that one problem with understanding the role of the family farm is that the family farm is not clearly defined. He attempts to clarify the problem by examining various definitions, presenting results of a structural analysis of agriculture, and considering some of the concepts used in the family farm debate. For example, Lemons shows that family farms can be defined as a family-controlled businesses, farms that are able to support a family, farms in which the land is owned by a family, or simply a small farm. Lemons writes that one of the difficulties of defining the family farm is that the definition may include as few as 5 percent of the nation's farms or as many as 95 percent, depending on the criteria for the definition. How the family farm is defined affects arguments about whether the farm is in decline, and whether research policy is partially or entirely responsible for that decline.

Gary Comstock also addresses the difficulty of defining the family farm in the collection of essays he edited entitled, *Is There A Moral Obligation to Save the Family Farm?* (1987). After a brief review of the difficulty, Comstock offers his definition, patterned after one proposed by author Luther Tweeten: "a family farm is an agricultural operation that is owned by a family or family corporation, has gross annual sales of between forty thousand dollars and two hundred thousand dollars per year, and does not hire more than 1.5 person-years of labor" (p. xxv). This definition accounted for less than 25 percent of the farms in the U.S. in 1985. Again, the variation in the definition can significantly alter an analysis of structural changes in agriculture.

Other authors consider the definition of the family farm in broader terms. For example, Deborah Fink explores some of the ideology of the family in "Constructing Rural Culture: Family and Land in Iowa" (1986). She concludes that, while the nuclear family unit was promoted in 19th-century agricultural development, evidence suggests that the extended family provided significant labor, as well as social and community needs. Furthermore, Fink argues that the traditional family farm ideology is no longer compatible with current goals of economic democracy. Instead, she suggests that "the experience of women on the land provides elements of an alternative ideology which stresses reciprocal nurture, cooperation, and shared good" (p. 43).

The value of women's contributions to the functioning of the family farm is the subject of several articles. Cornelia Flora, in "Women and Agriculture" (1985), reviews the roles of women in agriculture by examining the history of several cultures. She argues that the perceived role of women may vary from support to primary food producer, yet all cultures have relied on women to supply a significant portion of farm labor. In "Women's Work in the U.S.: Variations by Region" (1985), Carolyn Sachs investigates the role of women in maintaining the family farm. She argues that women's labor contributions, while often unrecognized or misunderstood, were important to production. Women often worked on a seasonal basis, providing key labor at harvest time. Another example is "A Commentary on Research on American Farm Women" (1985), by Peggy Ross. Ross presents an extensive review of research on women's participation in farming. Many of these articles illustrate the economic value of the labor that women supply, as well as the cultural value of agrarian ideals.

Jeffrey Burkhardt discusses the idea of a "farm crisis" in "Crisis, Argument, and Agriculture" (1988). He analyzes the use of the word "crisis" to describe contemporary agriculture and concludes that the loss of small farms does not in itself constitute a crisis. Concern about social values associated with "farming as a way of life" does, in his opinion, signal a larger social crisis involving our ability to sustain food and fiber production through political-economic institutions. Thus, Burkhardt attempts to separate structural definitions of the family farm from ideological ones and concludes that the current crisis is actually a crisis in the meaning of the family farm.

Ingolf Vogeler, in *The Myth of the Family Farm: Agribusiness Dominance of U.S. Agriculture* (1981), argues that the ideals of the family farm have been used to obscure the dominance of agribusinesses in agriculture. He identifies the primary myth as the belief that farmers are independent, when in fact they are controlled by impersonal economic and technological forces. Supporting myths include the work-ethic myth: Anyone who works hard will succeed; the free enterprise myth: The U.S. has a free-market economy in agriculture; the efficiency myth: Efficient farmers survive, while inefficient farmers go out of business; and the equal opportunity myth: Individual producers compete as equals in the marketplace. Vogeler claims that the USDA supports the myth of the family farm, because it "rests on an inadequate data base and on an inaccurate interpretation of the data" (p. 11). In other words, when factors such as indebtedness, off-farm employment, and contract farming are considered, the concept of an independent family farm becomes "illusionary" (p. 11). The implications of Vogeler's definitions on agricultural research policy are discussed in more detail in the following section.

Arguments

Gary Comstock's *Is There a Moral Obligation to Save the Family Farm?* (1987), is a compendium of viewpoints on family farms. Comstock borrows Kirkendall's classification of the "modernizers," who see "the present rural crisis not so much as a moral dilemma but rather as an unfortunate episode in the continuing saga of economic expansion and contraction," (p. xix) and "agrarianizers," who believe that farms are essential to our cultural identity, democracy, and virtues of independence, hard work, and family. The publication contains essays presenting arguments on such themes as history, the land-grant university complex, justice, and Christian theology.

Another useful introduction to issues in agriculture, including the family farm, is Jan Wojcik's *The Arguments of Agriculture: A Casebook in Contemporary Agricultural Controversy* (1989). Wojcik's method is to introduce each chapter with an anecdote or representative case, identify the issues, and review significant writers. Wojcik describes "sustainers" and "progressives," and concludes with an implied call for a movement away from agricultural research based on fossil fuels and chemicals. Some of the most emotional, and perhaps most widespread, arguments to save the family farm have centered on the argument that farming is more than just a business; it is a way of life with important moral, social, philosophical, and cultural effects.

Wendell Berry is perhaps the most eloquent voice in this group of authors. In his oft-quoted book, *The Unsettling of America: Culture and Agriculture* (1977), Berry describes the moral necessity of preserving the family farm because of the values embodied in that way of life. He argues that the colonization of America has always been a series of displacements of established people (native American Indians) in the name of progress. In his view, this pattern continues today with the farmers as victims, but without true progress. Instead, people have become separated from their communities and fragmented in their lives. Many have lost their work ethic and consumers are cursed with more material possessions and less happiness.

Berry's criticism of society in general and agriculture in particular is broad. He criticizes a false view of conservation which puts nature "at a distance, under glass." He believes that agriculture enhances character, work ethic, quality food, and the understanding of nature. This is the culture of agriculture that cannot be replaced by technology.

Berry disapproves of technology and the energy consumption required to sustain it. He suggests that the best agricultural technology would be diverse and lead to diversity, rather than the monoculture of agribusiness. It would encourage sustained farming, safeguard communities, preserve soil and water supplies, and encourage independent energy sources; in other words, it would nurture the land and people. Berry points to the Amish as an example of a community able to maintain a balance in their lives.

Other articles discuss quality of life on the family farm. For example, Thomas Auxter, in "Poetry and Self-Knowledge in Rural Life" (1985), argues that the changes in agriculture go beyond productivity gains or economic progress. To achieve self-knowledge requires an awareness of environment and nature. This awareness, articulated by writers such as Berry, places greater value on farming as a way of life. In "Agricultural Development and the Quality of Life: An Anthropological View" (1985), Peggy Bartlett and Peter Brown explore the relationship between agricultural development and quality of life by using anthropological concepts and examples. While most assume that agricultural development results in an improved quality of life, Bartlett and Brown illustrate that agricultural development is both harmful and beneficial.

Another group of articles considers the role of women in preserving the cultural values of the family farm. Joan Jensen writes, in "The Role of Farm Women in American History: Areas for Additional Research" (1985), that women traditionally were given the role of guarding and transmitting the values associated with farming as a way of life. Christina Gladwin also discusses the role of women in the agrarian myth in "Values and Goals of Florida Farm Women: Do They Help the Family Farm Survive?" (1985). Gladwin's research indicates that women value their work in terms of their belief in the agrarian way of life. Women show a commitment to the agrarian values of hard work, enterprise, and independence. They believe in keeping the family and the farm together.

Related to the cultural arguments are the environmental arguments. One important collection of essays devoted to environmental issues is *Agricultural Sustainability in a Changing World Order* (Douglas, 1984). Divided into three parts, the book discusses food sufficiency, stewardship, and community. Of particular interest to family farmers is the emphasis on stewardship of the land in relation to soil erosion, energy inputs, and pollution. Also important is the discussion of the human community and its need for healthy environments, rural cultures, and self-reliance. Authors contributing to this volume include Wendell Berry, Wes Jackson, John Cobb, Kenneth Farrell, and Vernon Ruttan.

Wes Jackson, educated as a biologist and geneticist, considers the problem of the family farm to be an environmental issue in *New Roots for Agriculture* (1980, 1985). He sees value in the cultural arguments of community and individual well-being, but emphasizes the effects of progressive, large-scale farming on soil erosion, energy expenditures, and chemical pollution. This short publication begins with a discussion of the failure of agriculture. Jackson argues that we have failed as stewards of the land, and our agricultural production records are also failures because of massive soil erosion and dependence on irrigation and chemicals. He claims we also have erred by ignoring the history of doomed civilizations that squandered resources. Finally, he says, we have failed because our national organizations, such as the Soil Conservation Service, have not put an end to the destruction.

From this indictment, Jackson suggests that a solution is possible— the development of perennials through plant breeding technology. Jackson believes that small farmers could enjoy greater yields with smaller expenditures on fertilizer, irrigation, and mechanical equipment. New production techniques would likely have small economies of scale. Finally, lower labor inputs would free farmers to pursue other interests. Jackson's writings also illustrate how environmental concerns are related to family farm issues.

Another collection of essays, edited by Wes Jackson, Wendell Berry, and Bruce Colman, is *Meeting the Expectations of the Land: Essays in Sustainable Agriculture and Stewardship* (1984). In the introduction, Jackson writes that the contributors to this volume believe a "radical change is necessary if we are to develop a sustainable agriculture" (p. xiii). Again, sustainable agriculture is related to the decline of the family farm. The essays discuss varied topics, such as the role of estate taxes; use of energy in farming; organic farming; anthropological and cultural tradition; and the impact of American attitudes and technology on Third World countries. Gene Logsdon's essay in this volume, "The Importance of Traditional Farming Practices for a Sustainable Modern Agriculture," argues for the use of animal power, varied crops and livestock, and limited chemical use as a method for providing a healthier environment for humans and animals, while maintaining an economically viable production system.

Many authors tend to minimize the sentimental or emotional arguments, focusing instead on farming as business. Jim Hightower's essay, "The Case for the Family Farm" (1975), summarizes the arguments as follows: small farms are more efficient; the price, taste, and nutritional value of food depend on family farmers; agricultural efficiency improved over the years because of family farms; family farmers, while being efficient, are being squeezed by both suppliers and processors; chronic low prices, lack of credit, and the market power of corporations hurt farmers; vertical integration takes control away from farmers; and the USDA and farm policy ignores the needs of small farmers.

Catherine Lerza and Michael Jacobson argue that food is power in their introduction to *Food for People, Not for Profit* (1975). They write that food is used by agribusinesses and large corporations for the generation of profits rather than the nutrition of people or the profit of individuals. As an example, Lerza and Jacobson cite canned baby formulas marketed in Third World countries where mothers often cannot afford to substitute formula for breast milk or do not have pure water supplies with which to mix the formula. They also decry the vertical integration of farmers, especially in the poultry industry, and "interlocking" arrangements between corporations, government agencies, and universities. Russell Parker and John Connor take a similar stance in "Consumer Loss Due to Monopoly in Food Manufacturing" (1979). They argue that any time a large market share is dominated by a few sellers, higher prices are certain in the long run.

Luther Tweeten reviews the moral philosophy used to judge performance of the food system in "Food for People and Profit" (1987a). He identifies two systems of food distribution: (1) the market system, which maintains that profit motivates people and competition is the best way to manage prices and supplies; and (2) the public sector system, which is a centrally planned and administered system. Tweeten concludes that the market system dominates, but has limitations which must be addressed. The public sector system supplements or corrects incentives and promotes stability and distributive justice. Elsewhere, Tweeten seems more concerned with economic policies than research policies. He sees larger economic forces at work in the demise of the family farm and concludes that research policy has a negligible effect on family farms. *Causes and Consequences of Structural Change in the Farming Industry* (Tweeten, 1984) considers some of these economic forces.

Greg Easterbrook challenges many prevailing ideas concerning the family farm in "Making Sense of Agriculture: A Revisionist Look at Farm Policy" (1985). He argues that "the actual circumstances of modern farming conform to few if any of the assumptions that underlie the public debate." Among the misconceptions that Easterbrook identifies are (1) farm families are not poor; (2) farmers are not being driven from their land; (3) farming is not a disastrous investment; (4) agribusiness does not dominate farming; and (5) most farmers do not have burdensome debt.

E. Wesley F. Peterson, in "Agricultural Structure and Economic Adjustment" (1986), reviews some of the factors contributing to changes in farm structure and compares these policies to policies adopted by the European community. He argues that the structural changes in agriculture (the movement from middle-sized farms to large farms) is an "example of economic adjustment in a capitalist economy" and policymakers should place these changes in the larger context of the general economy. Finally, Peterson argues that efforts should not focus on preventing change, but on helping farmers and communities adjust to changes.

Marty Strange also considers farm policies that have shaped American agriculture in *Family Farming: A New Economic Vision* (1988). He argues that a bias in federal commodity programs, income tax provisions, subsidized credit services, and technology all encourage the trend toward "bigness." Furthermore, he denounces the myth that farming efficiency improves with size, stating that policies encouraging bigness are counterproductive to the public welfare. He also has much to say about small-scale technologies, discussed in the next section.

Research Policy

History

Several important histories about the rise of agricultural science have been written throughout the years. Charles Rosenberg presents essays on the subject in *No Other Gods: On Science and American Social Thought* (1976). These essays represent a preliminary discussion of the relationship between social thought, social policy, and the prestige of science. Rosenberg discusses the development of the American scientific community, reviewing the lives of several pioneer scientists, their self-perceptions, and their roles in creating scientific institutions. This publication provides an interesting foreground to the problems facing agricultural research institutions today. As such, the essays illustrate the early struggles between applied and pure science, demonstrating the continued dilemma concerning scientists' perceptions of their research and the need to secure funding for research.

A recent work by Alan Marcus, *Agricultural Science and the Quest for Legitimacy* (1988), details a history of agricultural science and the events leading up to the enactment of the Hatch Act, which is responsible for the funding of agricultural experiment stations. More importantly, his work is a revisionist account of scientific and social conflict in agriculture in the 19th century. Marcus argues that agricultural experiment stations did not follow the European pattern (as previously assumed), but were the product of conflicting views of science—applied or basic, field or laboratory, etc. Ultimately, Marcus argues that the farm crisis is a social crisis, as well as a production crisis.

Christopher Dale, in "Agricultural Research as State Intervention" (1981), argues that agricultural research conducted or funded by the state benefits the "dominant class interests" of society. He uses Kentucky agricultural research experiment stations and the tobacco production industry as a case study. Dale traces the rise of the tobacco monopolies and their systematic attempts to create a vertical industry out of tobacco production at the turn of the century, thus squeezing out independent farmers. Then he reviews articles in Kentucky Agricultural Experiment Station publications during the same time period, which show support for increased production through technological innovation, and illustrate a marked absence of information addressing the economic problems faced by encroaching agribusiness monopolies. In his conclusion, Dale argues that the state "plays at least two key roles in capitalist society": (1) funding research to support dominant economic interests; and (2) providing legitimization for capitalist agribusiness via "scientific" research.

Criticisms of the Land-Grant University Complex

Many of the criticisms of the land-grant university complex come from the historical traditions described in the sections in this chapter on the family farm. These authors look at the current situation in agricultural research from a different perspective than the majority of agricultural research scientists. Because of differences in philosophical background, scientists and researchers need to understand the historical background of agrarian arguments to knowledgeably discuss agricultural research policy with agrarian supporters.

One of the earliest critics, Jim Hightower, wrote his most important book, *Hard Tomatoes, Hard Times*, in 1973. Hightower believes that a crisis situation in agriculture exists and that land-grant universities do not even recognize the crisis, much less provide leadership in the situation. Furthermore, Hightower believes that the system has abandoned its purpose as a "people's university," and through its alliances with corporate agriculture, has accelerated the crisis. Providing an historical overview of the acts that established the land-grant university complex, Hightower concludes that today's system violates the purpose of the original acts because farmers are no longer the primary focus of research or the product of the educational process. Hightower reviews research problem areas, specifically, funding allocations and scientific man-years; 85 percent of the funding was designated for technical research or production areas and only 15 percent for "people" areas, or projects designated to help rural communities, consumers, or rural Americans. He concludes that research allocations have led to a tax-paid research clinic for agribusiness. This bias in allocation of research funds, he says, encourages two types of research: (1) development of farm equipment (mechanization), which is best afforded by large corporate farms and hurts small farmers and farm laborers; and (2) development of new products, processing techniques, sales and marketing techniques, and vertical structures, which support agricultural input and output firms, but hurt small farmers and farm laborers. His criticisms derive from his belief in the value of the family farmer as an independent entrepreneur. Furthermore, Hightower believes the demise of the family farm is caused by monopolistic actions by agricultural researchers and agribusiness. His philosophical views are more clearly stated in his 1975 work, *Eat Your Heart Out: Food Profiteering in America*, in which he argues for the preservation of the family farm.

Wendell Berry also criticizes agricultural research policy and the failure of the land-grant universities to serve the people. Berry's comments are more philosophical than Hightower's. In Chapter 8 of *The Unsettling of America* (1977), Berry argues that Jefferson's ideal of a liberal education for a natural aristocracy was slightly altered by Morrill to be a practical or utilitarian education; that is, Morrill wanted to open the professional class of farming to laborers. In fact, in the Morrill Act of 1862, Morrill argues for the stabilization of farming communities and a permanent agriculture through education. Yet these ideals of education for farmers have degenerated until colleges of agriculture today promote impermanent agriculture and agribusiness graduates, not farmers. Another failure of the land-grant university complex, according to Berry, is not distinguishing agriculture from industry—a distinction found in the original language of the Morrill Act. Land-grant researchers have become caught up in furthering their careers, not in serving the public. Thus, Berry's implications for agricultural research policy revolve around his focus on the cultural integrity of farming as a way of life.

Ingolf Vogeler's assessment of the land-grant university system in Chapter 10 of his publication, *The Myth of the Family Farm: Agribusiness Dominance of U.S. Agriculture* (1981), continues Hightower's criticisms. Vogeler writes, "instead of focusing on the needs of rural *people*, the land-grant college system serves the interests of *capital* and *technology*—in short, agribusiness" (p. 199, his emphasis). He offers two cases to illustrate the consequences of technology on agriculture: mechanization of tomato production in California and mechanization of the Flue-cured tobacco industry in the Southeast. In both cases, Vogeler argues that the machines were developed using public funds in a land-grant university system and benefited large agribusiness concerns while forcing small farmers out of business and displacing farm labor. Vogeler also criticizes institutional racism in funding black land-grant schools and extension services. This lack of funding, encroaching technology, and the inability of black farmers to secure farm loans has produced a situation in which "black farm operators have found survival in southern agriculture more difficult than whites have" (p. 211). In summary, Vogeler believes farm workers and farmers are victims of technology because they have not been involved in agricultural research designed for their needs. These arguments, coupled with Vogeler's comments concerning the myth of the family farm, illustrate his belief that agricultural research policy is not responsive to the needs of rural Americans and communities, and much of the decline of rural America is because of bias in research policy.

Jack Doyle also shares Hightower's belief that scientists are not neutral in their decisions on allocations of funds for agricultural research. In *Altered Harvest: Agriculture, Genetics, and the Fate of the World's Food Supply* (1985), he argues that genetic technology and other high-powered technologies, contrary to the scientific wisdom which accepts these changes as empowering the U.S. to feed the world, have actually placed our food supply in far greater jeopardy than ever before. Citing the corn blight of 1970-71 as an example, he argues that overdependence on a few genetically engineered hybrid strains opens the possibility of future mutations, diseases, or pests capable of destroying food supplies. Doyle argues that the issue of commercial involvement in the university has been heightened with the advent of biotechnology. Scientists at universities are also involved with laboratory-based industries and often are preoccupied with applied research and product development, not basic research. The temptation is clear: Government funding of research has slacked off while industry funding has accelerated. A good university scientist can earn three times as much working for industry. Without industry funds, university laboratory equipment would be even more outdated than it already is. In a capitalistic society, researchers go where the money is. Doyle illustrates this concept of science for sale in two cases—one involving the petroleum industry and the other, pesticide use in agriculture.

In both cases, scientists were unwilling to testify about human health hazards because of the multiple conflict of interest or fear of future alienation from research funding sources. Such a situation in science calls for greater "public participation in the development of biological, agricultural, and biotechnological science policies" to ensure our future "power to feed a nation, and increasingly, much of the world" (p. 372). Doyle thus criticizes agricultural research policy, not from the direct standpoint of the family farm, but in terms of a more general denouncement of a biased research complex. He sees agricultural research as being unresponsive to public need or health, including the family farm.

Frederick Buttel responds to criticisms against the land-grant university system in his essay, "The Land-grant System: A Sociological Perspective on Value Conflicts and Ethical Issues" (1985). He begins his essay with a review of some of the important criticisms.

- The system develops technologies suitable for large farmers and agribusinesses.
- Agribusinesses, under pressure from world markets, criticize universities for failing to foster beneficial university-industry cooperation.
- A conflict exists between basic and applied research.
- Formula funding focuses on short-term results.
- The system has failed to consider the societal impacts of research.

After discussing value conflicts in land-grant research, extension, and teaching programs, Buttel challenges several of Hightower's assumptions, arguing that saving the family farm requires more than changing policies in land-grant systems.

David Danbom provides a readable and useful history in his essay, "Publicly Sponsored Agricultural Research in the United States from an Historical Perspective" (1986). He argues that even in times of decreased funding, agricultural researchers have maintained an emphasis on productivity, in spite of charges that they fail to achieve their goals and are indifferent to the needs of the people. Danbom also traces internal criticism, such as the National Research Council (NRC) report (1972) acknowledging environmental problems, another NRC report (1975) acknowledging environmental and social problems, and the 1981 Office of Technology Assessment (OTA) report citing research coordination problems and general organizational inefficiency.

Danbom suggests that while some of the criticisms are unreasonable, they are healthy because they call for a reevaluation of agricultural research policy. He also notes new areas of research to revitalize the field, such as nutrition, food and farmer safety, the environment, farm workers, and rural community development. Finally, Danbom cites Busch and Lacy (1983) as evidence that the system—because of its decentralization and variety—is capable of the flexibility needed to address many of these issues. His analysis provides insight about why criticisms have developed and offers hope that these criticisms can be beneficial for agricultural researchers.

Most of the writers who criticize the land-grant university system believe that it has been limited in its research focus and has ignored the historical goal of the system: serving the needs of the public. For many in agricultural research, these criticisms are neither realistic nor contemporary. Thus, a review of policy-making procedures is in order.

Policy Decisions and Their Effects

Agricultural research policy determines how scientists choose their research problems. While many believe that the choice is dependent on a need perceived by the scientist, in reality, scientists often must base their problem choice on availability of funds or goals set by administrators. In addition, factors such as how projects are approved, how funds are distributed, and how researchers are rewarded often play a significant role in problem choice. If the criticisms voiced by Hightower, Berry, Doyle and others are true—that researchers are biased in their choice of problems—then that bias should be addressed by researchers and administrators.

Research and the Family Farm (Carter et al., 1981) was written to address the problems of farm structure and the allocation of research funds. This report is the result of research conducted by a task force on behalf of the Agricultural Experiment Station Committee on Organization and Policy. The authors begin with an historical overview, showing the decline of the small farm and the concentration of farm resources among large farms. While the authors offer several possible explanations for the decline—availability of high off-farm income, for example—they admit that "technological advances have fueled the treadmill and absorption processes" (p. 5). This results in a decline in the numbers of small farms and a concentration of resources for large farms. Furthermore, the authors show that agricultural experiment stations were a part of the development process of technology.

But what are the effects of research funding allocation on farm structure? A random survey of agricultural experiment station funding for 1979 shows that 74.3 percent of the sample was designated for research for agricultural production; 10.9 percent for marketing and processing; 7.3 percent for family living; 6.0 percent for community; and 1.5 percent for other projects. Of the agricultural production research, 28.1 percent was on basic research; 3.3 percent on public bodies; 7.3 percent on small farms; 4.9 percent on medium-sized farms; 3.2 percent on large farms; and 53.3 percent was size-neutral. The authors emphasize that their survey was a sample of agricultural experiment station funding only; it did not include all public and private research. Nonetheless, they conclude that, at least for agricultural experiment stations, research is not slanted toward large-farm concerns. Instead, the authors contend that other economic concerns, such as government policies, have a greater impact on farm structure.

The authors offer the following proposals for research strategy:

- Scientists should emphasize size-neutral research. The needs of the consumer for cheap, abundant food often conflict with the needs of the producer and concerns for the environment, conservation, etc. Nevertheless, these conflicts should be addressed through economic and government policies because research has minimal effect on farm structure.
- Medium-sized farms often have trouble with cash flow and high capital requirements. Research can play a role in providing better management techniques and expertise in financial, marketing, and production management.
- Small farms include a diverse group of old/young farmers, full-time/part-time farmers, etc. The research emphasis should not be on small farms, but on the rural poor. Research in poverty and underemployment problems is needed.
- Finally, the authors recommend continued, aggressive research on agricultural productivity, with emphasis on biological and resource management concerns as well as policy development, rural development, and economic agricultural issues, such as tax and credit policies.

Edited by Lawrence Busch, *Science and Agricultural Development* (1981), is a collection of essays which originated in a graduate seminar focusing on the role of science in agricultural development. Like the authors of *Research and the Family Farm*, Busch also recognizes a problem in agricultural research policy and identifies it as an emphasis on efficiency and short-term profits. Furthermore, he notes that the funding for agricultural research was stable until the mid-1970s. Since then, appropriations have declined and the assumptions of agricultural research have been questioned. Without continued research into environmental issues, the social and economic aspects of agriculture, and agricultural policy, however, we risk our food supply for the upcoming century. Thus, Busch evaluates the role that science played in agricultural development and its effects on farm structure and society in general.

Part I of the publication (Busch, 1981) explores the relationship between the agricultural sciences and capitalism, and includes an essay on the history of the mechanization of agriculture in the 19th and 20th centuries. Essays in Part II, "Science, Scientist, State: The Evolution of Agricultural Technology," consider the context and assumptions of agricultural research policy. Finally, Part III addresses the exportation of agricultural research and development to Third World countries. It offers alternatives to agricultural productivity research which has heretofore dominated agricultural research.

In the same collection, Lawrence Busch and William Lacy examine the influences on scientists in choosing areas of research in "Sources of Influence on Problem Choice in the Agricultural Sciences: The New Atlantis Revisited." Beginning with an overview of external and internal sources, the authors state two positions: (1) the internalist position, which argues that scientists and science are autonomous and use objective techniques for studying; and (2) the externalist position, which argues that science is simply a product of the social position and merely legitimizes class interests. To examine both internal and external influences on the choice of problems, the researchers conducted interviews with informants and reviewed the publications of agricultural science. They conclude that internal influences—the process by which researchers are socialized (graduate school programs), postgraduate training programs, new theoretical orientations and methodologies, the historical development of agricultural disciplines, and discipline boundaries established by faculty selection at universities—all serve to influence problem choice in research.

Among the external influences, the commodity groups that financially and politically support agricultural research had the greatest influence, along with social movements, consumer concerns about health and environment, the changing economic context, and the availability of research funds. Another consideration is institutional pressure to publish, which leads many researchers away from long-term projects. Their work thus reinforces the criticisms of some writers that agricultural research scientists are not responsive to alternative research areas such as social and environmental concerns.

In *Science, Agriculture, and the Politics of Research* (1983), Busch and Lacy continue their discussion of agricultural research policy, reporting the results of a detailed study to examine the factors influencing the choice of research problems by agricultural researchers and scientists. The authors recognize that productivity is at a peak, but question the ability of agriculture to sustain these limits in light of the fragility of soils, water systems, and pollution. The authors also point to the fragmentation of the farm population as an area for concern. Given this situation, how do scientists choose research problems?

Busch and Lacy answer that question with a more complete account of the study and results already discussed. In addition, they address the problems inherent in the political context of agricultural research. The majority of funds is provided by the government and is thus influenced by various interest groups. Because of this political reality, they argue that some criticisms by Hightower (1973) and Berry (1977) may be justified. But they maintain that without "active cultivation of support, agricultural research would cease to exist" (p. 235). In fact, many of the traditional sources of support are declining, while new interest groups such as organic farmers, consumers, environmentalists, and farm workers are emerging. Busch and Lacy suggest that this change can both harm agricultural research if scientists squabble among themselves or benefit research by providing new sources of funding and revitalization.

In relation to the need for change, Busch and Lacy do not desire a revolutionary approach to agricultural research policy, but do offer the following "modest changes" (p. 237):

- A national organization is needed to speak for all agricultural research.
- More interdisciplinary communication and research is needed through journals, graduate school programs, and societies.
- Hiring procedures should encourage breadth of knowledge as well as depth, and qualified women and minorities need to be given research opportunities.
- Research agencies and advisory groups must recognize a broader range of issues and client concerns and high-risk, long-term research should be encouraged.

- A national agricultural research policy is needed that can be integrated into an overall national agricultural policy.

Each of these suggestions, if implemented, could have an effect on
family farms and farm structure by recognizing that qualities other
than productivity are justifiable research goals.

Kenneth Dahlberg provides a collection of essays that consider the
ethical and value assumptions of agriculture as well as the "neglected
aspects of agriculture"—social, health, resource, and environmental
concerns—in *New Directions for Agriculture and Agricultural Research:
Neglected Dimensions and Emerging Alternatives* (1986). This book, which
evolved from a project funded by the Ethics and Values in Science and
Technology Program of the National Science Foundation and the National Endowment for the Humanities, brings together information on
ethics, global resources, historical overviews, and interactions between
agriculture and society.

In the introduction, Dahlberg reviews the major issues affecting agriculture today. He argues that changes during the last couple of decades
have made basic resources such as land, water, energy, labor, and
capital more expensive. Another important change is the way people
view their environment and the recognition of the dangers of uncon-
trolled pesticides and chemicals. Finally, economic issues, such as
distribution of resources and the social effects of mechanization, are
also emerging as significant changes in the agricultural context. These
changes call for a review—and if necessary a change—in agricultural
research policy.

Dahlberg notes that the awareness of a need to review agricultural
research policy is already underway. He cites Hightower (1973), Berry
(1977), and Busch (1981) as calling for a change in priorities, albeit with
divergent remedies. His goal in this collection, then, is to offer the
background needed to understand the changing context of agriculture
as well as to consider alternatives. Dahlberg also provides an evaluative
approach in considering the underlying values and assumptions of
researchers and policy-makers. This evaluative approach is necessary,
he argues, because intelligent and educated people often disagree sharply
on policy issues. Only by considering their underlying assumptions
can they hope to understand each other's perspectives. These essays
thus serve as an introduction and discussion of recent changes in
agriculture, including the increasing emphasis on family farms.

Several other essays correspond to Dahlberg's discussion of the need
to consider value assumptions. They point to a disparity in the arguments of goals for research as a difference in values. Rachelle Hollander,
"Values and Making Decisions about Agricultural Research" (1986),
uses four value presumptions developed by William Aiken to consider

how policy disputes arise and may be reconciled. She identifies the four normative positions as (1) top priority view, (2) net gain view (3) constraints view, and (4) holistic view. Next, she uses several recent controversies in agricultural research to illustrate that two individuals may argue with facts, but if they argue from different value positions, their arguments cannot be reconciled. Thus, she maintains that many disputes about policies and practices in agricultural research are actually disputes about values. She suggests that researchers should consider these value differences in the decision-making process about policy matters.

Cornelia Butler Flora also addresses values in agricultural research policies in "Values and the Agricultural Crisis: Differential Problems, Solutions, and Value Constraints" (1986). Flora argues that people have assumed that "a single policy instrument will solve all problems" of the farm crisis (p. 17). She demonstrates that many groups are involved in the crisis—farmers, farm families, rural communities, banks— and each group has different needs. Two of the most important assumptions preventing resolution are the anti-welfare attitude and anti-planning attitude. Finally, Flora argues that "we cannot begin to address the wide variety of problems in combination that constitute 'the farm crisis' until we legitimize the rightness of defining the kind of agriculture the United States wants" (p. 22).

In a related discussion, Paul Thompson argues that utilitarian ethics provide a model for evaluating moral responsibility in agricultural research decisions, thus balancing the costs and benefits for the public, in his essay, "Ethics in Agricultural Research" (1988b). Yet utilitarian ethics is not without its deficiencies. For example, it does not consider the inequitable distribution of benefits and harms or anticipate all possible benefits and harms. Future researchers considering the moral responsibility of agricultural research may profit from a model that includes consideration of the deficiencies of utilitarian ethics.

The essays discussed in this section suggest changes in agriculture which call for a reevaluation of agricultural research policy. Authors may debate to what extent research policy has affected or hastened the decline of the family farm, but all agree that changes in population, environment, and the economy call for another look. Furthermore, Dahlberg, Flora, Hollander, and Thompson demonstrate the problem of value assumptions in evaluating agricultural research policy. They maintain that understanding the historical development of ideas about agriculture is a prerequisite to understanding the discrepancy in many of the arguments today. Once these discrepancies are clarified, agricultural researchers can respond to the criticisms, and if appropriate, integrate them into their research policies.

Role of Technology

While agricultural research incorporates many facets of agriculture, the development of technology has been identified as one facet with important consequences for family farms. Several essays consider the role of technology in maintaining or undermining family farms and offer some implications for future developments and effects.

One of the earliest researchers to discuss the effects of technology and economies of scale was John M. Brewster in *A Philosopher Among Economists*. This collection of essays, edited by J. Patrick Madden and David Brewster (1970), was prepared from papers written in the 1950s and 1960s and left unpublished at Brewster's death in 1965. Brewster shows that Americans have overcome national problems of hunger, first through frontier expansion in the 19th century and secondly through technological advances in productivity. That same productivity which stabilized our population, however, established such an accelerated pace that many farms eventually went out of business. In a related essay, Brewster discusses technological innovations that could eventually transform the sequential pattern of agriculture into a simultaneous pattern. Each of these innovations has important effects on people, however, and his emphasis on the role of the social scientist in the development of policy- and decision-making sounds quite contemporary.

Harold Breimyer explores the effects of partial industrialization and farm policy on agriculture in *Farm Policy: Thirteen Essays* (1977). In the third essay, Breimyer argues that technology has the ability to relieve human labor and improve the productivity of land and animals. Yet these changes are not without questionable effects, such as freeing farmers for other employment and industrial regimentation. In the same essay, Breimyer maintains that today's agrarian features of agriculture tend to maintain smaller, independent farms, while the industrial features of agriculture move toward larger farms. The future structure of agriculture is unknown at this time, according to Breimyer, yet he advocates technically oriented research and education for farmers through agricultural experiment stations and extension services. Breimyer does not, however, address the inequalities of research or the dissemination of research expressed by Hightower and others.

A significant report by the Office of Technology Assessment (OTA), *Technology, Public Policy, and the Changing Structure of American Agriculture* (1986), projects changes in agriculture for the next 15 years. In the past, public research was an important factor in developing new technologies—both mechanical and biological—to improve agricultural production. While the public sector has continued to make monies available for research, the report states, dissemination of research through extension services has failed in recent years. The large agribusinesses and farming corporations have kept abreast of changes in technology because of private research and consultants.

Following a detailed discussion of these changes, the OTA suggests several policy changes:

- Reevaluate the social contract for the agricultural research and extension system. New guidelines should be developed that encourage the system to protect the public interests.
- Maintain a leadership role of public research at state and national levels in spite of increases in private research.
- Maintain local or regional research centers because of their valuable contributions to various agro-ecosystems, and eliminate the dichotomy of adequately and inadequately funded universities that has developed in the past.
- Maintain the role of the extension service in areas such as product testing and evaluation.
- Improve special funding opportunities for the transfer of technology to medium-sized farms because "while agricultural research is not inherently biased toward large-scale farms, lags in adoption by small and medium-sized farms have the same effect" (p. 279).
- Provide funding for extension staff development to ensure that specialists and county agents keep pace with technology changes.
- Reevaluate the roles of extension specialists and county agents in the agricultural research system because these roles have changed and the current structure no longer adequately serves the needs of medium-sized and small farmers.

This report implies that much can be done to make technology available for small and medium-sized farms. Without these policy changes, however, the OTA suggests that recent trends will continue: The number of small farms will continue to decline, and an increasing concentration of productivity will be devoted to large farms.

Most technology discussions focus on technological change or innovation and the effects on agricultural production. A collection of essays by Gigi Berardi and Charles Geisler, *The Social Consequences and Challenges of New Agricultural Technologies* (1984), examines the social and cultural implications of technological change. The essays are grouped into historical, recent, and emerging trends. The historical accounts consider the impact of technology on labor displacement and related issues such as new employment for displaced workers and government benefits. Five articles which address the development of machines to harvest products in California seek to determine the social costs and benefits associated with the mechanical tomato harvester. Other essays consider lettuce, grape, and citrus production. The final section focuses on contemporary issues and illustrates the complexity of evaluating technological impact on social, economic, and psychological concerns. This publication represents an important new trend in agricultural research: an awareness of the social and economic effects of technology. Furthermore, it demonstrates an attempt to evaluate the benefits of technology, not only on the basis of increased productivity as in the past, but also on broader issues of justice and the distribution of benefits.

In a 1987 essay, "Ethical Perspectives on Changing Agricultural Technology in the United States," Patrick Madden and Paul Thompson also consider the benefits of technology in terms of justice and equitable distribution of benefits. They argue that agricultural technology has caused three major effects:

1. Food production and agricultural productivity have increased.
2. Farm structure in the U.S. has shifted from many medium-sized farms to a few large farms that now produce the majority of food, and many small, part-time farms.
3. Many new environmental questions concerning water pollution and food quality have emerged because of the use of chemicals. Each area is evaluated in terms of the ethical implications, resulting from changes in technology, on farmers, agribusinesses, farm workers, and consumers.

First, Madden and Thompson show that technology changes in food production generally lead to fewer hours of human labor and more capital-intensive farming techniques. These changes also lead to displaced farm workers (hired labor) and many less efficient or slow-to-adapt farms that go out of business. Using mechanical harvesters and the bovine growth hormone as case studies, the authors show that a market readjustment following the introduction of a new technology does not always lead to improved profit margins for farmers. Also, the negative effects of the technology often outweigh the positive benefits of lower consumer costs (i.e., government price supports).

Madden and Thompson also review the arguments of Hightower and Berry but find only a weak argument for agrarian themes. If family farms are crucial to a national self-concept, sufficient reasons may exist for preserving this way of life and opposing the technology that threatens it. The authors' recommendations for future research and consideration stress environmental, rather than agrarian themes. Their suggestions reemphasize the call by Busch and Lacy and Dahlberg to focus on agricultural research issues other than productivity, but do not necessarily reflect an overriding concern about agrarian themes.

Conclusion

As the organization of this chapter illustrates, no real consolidation of information on farm structure and agricultural research policy exists. This divergence is evident from both the content of the literature surveyed and the indexes used to research the fields. The articles and books addressed the subjects separately. Rarely was the alternate topic given more than cursory treatment or even listed in the bibliography. Likewise, computer databases used to index articles, books, and government reports lacked access to the subject. Key words and synonyms could access family farm/farm structure or agricultural research policy, but when the two subjects were combined, the databases listed fewer than 10 citations. The problems in researching the field indicate that the combined subject—how perceptions of the family farm affect the development of agricultural research policy and, conversely, how agricultural research policy impacts farm structure—is not yet recognized as a valid research area. This chapter demonstrates the feasibility of integrating these two areas and suggests the need for additional research.

10

Environmental Goals in Agricultural Science

Paul B. Thompson, Gary E. Varner, and Deborah A. Tolman

Introduction

Critics of agriculture and agricultural research often question the effort to improve productivity and efficiency in farm production with respect to impacts upon environmental quality and the animals involved in dairy, beef, and poultry operations. Criticism of yield-enhancing agricultural research can be based upon the belief that alleged efficiencies are false when long-term resource costs are included in a factor analysis. Alternatively, the environmentalist may believe that destruction of habitat is always wrong, without regard to production economies, and the animal rights activist may believe that suffering should be minimized wherever it occurs (whether in human or non-human animals), even when it is not cost-effective from a production standpoint. The alternative arguments may be based on the judgment that non-human animals (and even ecosystems) have moral value or rights similar to those of humans. The goal of this chapter is to identify and differentiate some of the main themes in and implications of the ethical judgments that underlie the environmentalist and animal rights criticisms of agriculture and agricultural research.

The literature on the environmental impacts of agriculture is already quite extensive, and is expanding at an increasing rate. Although this chapter is designed primarily as a bibliographic essay, it will make no attempt to survey that literature and, indeed, will omit many of the most important works by environmentalists, ecologists, and agricultural scientists. A new selection of literature on sustainable agriculture, for example, has emerged prominently in the past decade, culminating in the National Research Council's 1989 report *Alternative Agriculture*. The discussion of sustainable agriculture has become embroiled in philosophical debate about definitions.   For example, Brundtland's (1988) approach to sustainability defines the issue in terms of intergenerational equity, a "poor people's" approach, while Pearce et al. (1987) contends that "maximizing net benefits of economic development, subject to maintaining the services and quality of natural resources over time" should be the focus. While a survey of this literature would be useful, it would complicate and detract from the primary mission of this chapter.

The comprehensive goal of this book is to examine the role of ethical concepts and values in planning, conducting, and disseminating the results of agricultural research. In this context, environmentally based criticism of agriculture is important because critics have sometimes expressed or insinuated the opinion that ethical obligations to protect environmental quality and respect the interests of farm animals have been neglected. There are many ways in which such a judgment might be conceptualized in ethical terms, however, and the purpose of this chapter is to survey some of the main strategies that critics have used to justify their views.   Relatively few authors make any attempt to demonstrate how environmental goals are linked to ethical responsibilities or to the historical mission of the USDA/land-grant system. Although many citations describe the environmental impacts of agriculture, agricultural research policy, animal welfare, and environmental ethics, very few of them discuss how these four topics are interrelated. The bibliographic notes interpret how the linkages might be connected, even when authors do not specifically address all three topics in the sources cited.

Because the primary audience for this essay is agricultural professionals, we will concentrate primarily on ethics citations, the literature with which our audience is least likely to have prior knowledge. However, when discussing the environmental ethics literature, the emphasis will be upon applications to agricultural production research issues. Topics such as the protection of endangered species, or conflicts between recreation and wilderness preservation—topics that have received significant attention from environmental ethicists—are not discussed here. When possible, the focus will be narrowed to issues that impinge directly upon agricultural research.

Environmental Criticisms of Agriculture

Although we soon will summarize the key themes in environmental ethics, we first discuss the ethical significance of environmental criticisms of agriculture for the traditional ethic of the USDA/land-grant system. The introduction of this book outlines the utilitarian character of traditional agricultural research planning. The implicit ethical basis for agricultural research in the USDA/land-grant system is *utilitarian* in that it defines the value of research in terms of its capacity to balance costs and benefits associated with agricultural production. It is also *anthropocentric* insofar as the balancing act is limited to the costs and benefits to human beings. The most justifiable research project is the one that promises to achieve the greatest good for the greatest number of people. When costs to non-human animals and the environment are ignored, the utilitarian view of social ethics makes it easy to think of ethical evaluation as a form of calculation in which all the benefits and costs of various options are weighed. From this perspective, the ethically justified course of action is the one which best satisfies the rule of maximizing good (or in situations where all options are unattractive, of minimizing evil).

Even when costs to non-human animals and the environment are bracketed, however, one general problem remains with optimizing decision procedures such as the utilitarian ethic just described: It is impossible in practice to obtain complete and reliable information on all the relevant consequences of a policy decision. Some factors are inevitably left out, and, when these factors affect human health and safety, the economic well-being of minority groups, or the quality of the human environment, the entire moral calculation of relative benefits can be drastically modified. Costs or harms that are simply left out of a utilitarian calculation are called *externalities*. Costs are sometimes left out because the decision-maker does not have a reliable way to measure or compare them, and costs also are left out when individuals or groups deciding on behalf of their own interests do not have to bear them. For self-interested decisions, the costs are truly external to the decision-maker's point of view. Decision-makers entrusted with the public good must make stringent efforts to reflect all such "externalities" in any estimate of social benefit or harm, or they cannot truly have optimized outcomes.

The majority of environmentally based criticisms of agriculture and agricultural research clearly take the form of noting externalities, or factors and impacts that have simply been left out of the assessment of costs and benefits. The most celebrated of all works in the critical literature, Rachael Carson's *Silent Spring* (1962), follows the pattern of citing externalities by providing an extensive list of unintended consequences associated with the use of insecticides. Carson was a gifted writer with a flair for evoking the beauty and dignity of wildlife and natural habitat. Nevertheless, her basic philosophical strategy in the book was simply to identify unintended consequences. *Silent Spring* is a laundry list of externalities that had not been accounted for in making assessments of pesticide use in agricultural production.

Silent Spring is just the first in a long line of critiques identifying environmental externalities of chemicals used in agricultural production. Carson's arguments were reiterated and extended in Frank Graham's 1970 book *Since Silent Spring* and in Robert Van den Bosch's *The Pesticide Conspiracy* (1978), to cite two of the most prominent titles. While Carson noted toxic effects upon non-target species, Van den Bosch identified unintended consequences that typify the class of ecosystem outcomes of intense interest to environmentalists. According to Van den Bosch, chemical pest control ignores ecological forces that control insects. The number of insect species classified as pests doubled from 1962 to 1978, despite increasingly efficient chemical control, skyrocketing insect control costs, and increasing environmental impacts, a phenomenon that Van den Bosch labels an "insecticide treadmill." Insecticides or biocides kill natural enemies of insect pests, and eradicate the natural predators and parasites. A biotic vacuum is then created where the surviving pests, without predators or parasites, thrive. Frequent insecticide sprayings have unintended consequences because resurgence of the target pest occurs, and minor pest species become major problems when their natural enemies are eliminated by chemical sprays. Continued spraying then becomes a necessity.

Van den Bosch makes an explicit ethical argument when he indicts chemical company salespeople and advertisers for their effect upon a farmer's decision to use more pesticides and contribute to the treadmill. Scientific societies and administrators of land-grant universities, where the nation's pest control research is conducted, are also implicated when pressures and political reprisals from chemical companies affect USDA/land-grant research programs. Van den Bosch's condemnation of public sector agricultural research is based upon a judgment that administrators and scientists have been "captured" by the commercial pesticide industry, and thus have failed in an ethical obligation to conduct research in the public interest. His criticism is still consistent with Carson's original pattern of identifying externalities, however, for Van den Bosch clearly understands the public interest in terms of maximizing benefits for human beings. The problem he has with chemical insecticides is that their costs outweigh their benefits. It is profitable for chemical companies to sell pesticides and for farmers to use them only as long as important long-term costs are not included in the overall assessment (in other words, as long as the costs are "externalized"). In this case, costs are externalized either in the sense that they are borne by individuals whose interests are not included in the tabulation of consequences or in the sense that costs occur beyond the time frame for which consequences have been assessed.

John Perkins identifies an additional form of externality in *Insects, Experts, and the Insecticide Crisis: The Quest for New Pest Management Strategies* (1982). He compares the consequences of integrated pest management (IPM) and total pest management (TPM) crop management strategies by claiming they evolved in response to problems with chemical control. Both use principles of population dynamics to suppress and/or manipulate pest populations. TPM produces ecosystem-wide effects, irrespective of boundaries, in that eradication measures are taken against major pest species. IPM does not include eradication as a goal, and, hence, impacts are confined in both space and time. The point is that the systemic impacts of TPM necessarily produce externalities in virtue of the fact that TPM produces inherently unpredictable changes in the basic ecology of natural systems. As such, it becomes impossible to complete the assessment of trade-offs from TPM, and this, in itself, provides an edge for IPM. The ethical basis of Perkins' work on insecticide policy is reviewed by Rabb (1984).

The prolific writings of David Pimentel, working with a variety of co-authors, document a series of environmental externalities associated with agricultural production. With Carl Hall, in the recent work *Food and Natural Resources* (1989), Pimentel proposes an ethical standard of maximizing the benefits of biotechnology while reducing the risks involved. He concludes that difficulties in policy-making are compounded by the complex nature of the environment and population, but an environmental policy must include a sustainable agriculture, one that remains for future generations of humans. Both Pimentel and Hall suggest that economic values of stewardship and utilitarianism must continue to shape our agricultural system for us to continue to obtain food and other resources from the land. Pimentel objects to the belief that human beings may have an unlimited right to manipulate environmental resources for personal gratification. The basic pattern of Pimentel's argument, however, is to accept the traditional utilitarian basis of agricultural research planning, but to insist that research planners do a better job of including neglected consequences in the assessment of agricultural practices.

A substantial amount of literature exists on how to do a better job of including externalities in the assessment of agricultural production practices and agricultural research. Barkley and Seckler (1972) is an early, but still relevant and readable introduction. Alan Randall (1981) provides a textbook discussion of techniques for including (or internalizing) a wide variety of environmental values into an assessment of basic policy choices. These techniques are as applicable to agricultural research decisions as they are to familiar forms of environmental regulation at the federal level. It is important to emphasize the extensive theoretical and empirical development of these techniques, and to lament (with Vernon Ruttan, this volume) that they are not more widely utilized. Recent discussions can be found in Smith (1990), which has an extensive list of references, and Johnson and Johnson (1990). Nevertheless, the balance of this chapter will address philosophical criticisms of agriculture that follow a substantially different direction.

Founding Fathers of Environmental Ethics

Although *Silent Spring* probably marks the beginnings of environmentalism as a movement to curtail industrialization through the regulation of technology, the philosophical foundations of environmental ethics refer back to the work of John Muir, founder of the Sierra Club, and Gifford Pinchot, founder of the National Forest Service. Muir and Pinchot represent two poles of thought that define the fundamental tension in environmental philosophy. Muir was the advocate of nature for nature's sake, of preserving vast tracts untouched by human hands. Pinchot was the advocate of managing resources for long-term productivity and future use. Pinchot was, in many respects, a paradigm example of the enlightened utilitarian, sensitive to the problems of externalities that have, today, become the main focus of resource economics. Muir, on the other hand, believed that it was inappropriate to calculate the value of protecting natural resources by anticipating the costs and benefits for human populations. Muir's discomfort with Pinchot's enlightened utilitarianism marks the course for many of the most severe environmental critics of agriculture.

Muir's heirs see unwanted impacts upon environmental quality as a symptom of a shallow and arrogant understanding of humanity's place in nature, rather than something approximating an economic externality. These critics frequently reject economizing or optimizing policy strategies as incapable of adequately comprehending the interactive dependence of human populations and natural systems. They argue that goals must be derived from an understanding of ecology, and that natural systems place constraints upon agriculture which have been neglected in the past. One of the first attempts at a systematic intellectual account of this tradition was made by ecologist and wildlife manager Aldo Leopold, who began his career as a disciple of Pinchot. Leopold's intellectual history, including his break with Pinchot's utilitarianism, is recounted in Norton's chapter of this book, as well as in recent studies by Flader (1974), Callicott (1987), and Nash (1982, chapter 11; and 1989, chapter 3).

Leopold is best known among philosophers for his essay, "The Land Ethic," from *A Sand County Almanac* (1949). The essay begins with a passage in which Leopold describes the rejection of human slavery as one of the key instances of moral progress in history. The key to this event, he thinks, was ceasing to understand human beings as property, or extending the scope of the moral community to include all human beings (Leopold, 1949, pp. 201-203). Leopold's message is that we must now find a way to think of our relationship to the land (meaning the biosphere in general) as something other than mere property.

Leopold believes that attempts to reflect conservationist concerns within the kind of optimizing calculations that underlie a traditional approach to agricultural decision-making are hopelessly lacking. In Leopold's view, ample basis exists for care and concern about ecological values, but the problem is that the importance people place upon nature cannot be reflected in monetary terms. He writes:

> When one of these non-economic categories is threatened, and if we happen to love it, we invent subterfuges to give it economic importance. At the beginning of the century songbirds were supposed to be disappearing. Ornithologists jumped to rescue with some shaky evidence to the effect that insects would eat us up if birds failed to control them. The evidence had to be economic in order to be valid. (Leopold, 1949, p. 210).

Here Leopold also seems to be rejecting the notion that unwanted outcomes of agricultural production decisions can be accommodated by a broader framework of costs, including some constraints. Indeed, property rights, Leopold's target, serve as the model for constraints. Instead, we must rethink our lives to attain a fuller appreciation of the interdependence between human and natural communities.

One of the chief sources for understanding these connections is agriculture. Early on in his book Leopold writes, "There are two spiritual dangers in not owning a farm. One is the danger of supposing that breakfast comes from the grocery, and the other that heat comes from a furnace" (Leopold, 1949, p 6). One who lives on a farm cannot, in Leopold's view, long forget the dependence of human action upon the underlying natural ecology. Written in the 1940s, *A Sand County Almanac* does not reflect the more recent critics' concern that agriculture is on the verge of destroying its ecological base, but Leopold does express cynicism about the optimizing strategies of experiment station research:

> The State College tells farmers that Chinese elms do not clog screens,

and are hence preferable to cottonwoods. It also pontificates on cherry preserves, Bang's disease, hybrid corn, and beautifying the farm home. The only thing it does not know about farms is where they came from. Its job is to make Illinois safe for soybeans (Leopold, 1949, p. 117).

Leopold's land ethic, thus, rejects the optimizing strategy that takes increasing income, increasing production, and increasing benefits to consumers as its core. Instead, Leopold urges us to:

Examine each question in terms of what is ethically and esthetically right, as well as what is economically expedient. A thing is right when it tends to preserve the integrity, stability, and beauty of the biotic community. It is wrong when it tends otherwise (Leopold 1949, pp. 224-25).

E. F. Schumacher's 1972 book *Small Is Beautiful* followed *Silent Spring* by a decade, and combined Carson's concern for agricultural technology with Leopold's distaste for making moral evaluations by calculating costs and benefits. The central theme of the book was widely taken to be an attack upon technologies that consumed relatively large quantities of fossil fuels and required large investments of fixed capital. In the chapter entitled, "The Proper Use of Land," however, Schumacher addresses a central question in agricultural ethics. The argument of the chapter is first a criticism of what Schumacher calls "the philosophy of the townsman," and second, a description of an alternative program. The "townsmen" see agriculture's economic woes as evidence that farming or ranching is a declining enterprise, and view the central problem of agriculture as one of improving farm income. Schumacher finds this view deficient. He writes:

We know too much about ecology today to have any excuse for the many abuses that are currently going on in the management of the land, in the management of animals, in food storage, food processing, and in heedless urbanization. If we permit them, this is not due to poverty, as if we could not afford to stop them; it is due to the fact that, as a society, we have no firm basis of belief in any meta-economic values, and when there is no such belief the economic calculus takes over (Schumacher, 1973, p. 116).

In Schumacher's opinion, the problem arises when agriculture is understood as essentially defined by its capacity to produce and market salable commodities. In making a statement of the wider goals of agriculture he writes the following.

A wider view sees agriculture as having to fulfill at least three tasks:
- to keep man in touch with living nature, of which he is and remains a highly vulnerable part;
- to humanize and ennoble man's wider habitat; and
- to bring forth the foodstuffs and other materials which are needed for a becoming life.

I do not believe that a civilization which recognizes only the third of these tasks, and which pursues it with such ruthlessness and violence that the other two tasks are not merely neglected but systematically counteracted, has any chance of long-term survival (Schumacher, 1973, p. 113).

These remarks on agriculture must be understood in the light of Schumacher's overall attack upon "economic values," and his campaign to substitute a norm of "Buddhist economics" in its place. In criticizing economic values, Schumacher means to attack the utilitarian emphasis upon increasing incomes; by interposing "Buddhist economics" in place of this emphasis, he intends to suggest that there is an alternative way to conceptualize economic activity, one that would trace production, distribution, and exchange according to the long-term impact of these activities upon the natural systems needed to support all. Economic policies that encourage consumption in order to promote economic growth are, in the view of Buddhist economics, incompatible with the goal of a permanent and stable society (Schumacher, 1973, pp 30-33).

Although Schumacher's choice of words has the ring of late-sixties hippie jargon, his point should be understood as a shift in philosophical perspective. Political theorist Paul Diesing has argued that Schumacher's critique is a complete rejection of the traditional utilitarian perspective on agricultural production. Reading this traditional view, Diesing writes:

[N]ature appears in three forms: natural resources, cultivated land ... and externalities of production. Natural resources are free goods, *res nullius*, nothings, having no value until they are "produced" and made available for exchange. (Diesing, 1982, p 294)

When the central goal of agriculture is understood in terms of production, agricultural land is a form of fixed capital, and this, in turn, suggests that this land should be devoted to its most productive use. Although unwanted outcomes can be factored into the optimizing equation either as costs or as constraints, the result looks a bit like pre-Copernican models of the solar system, where epicycles and reversing

rotations were added onto the charts for planetary motion to preserve a
theory that falsely placed the Earth at the center of the universe. In
Diesing's view, Schumacher rejects this strategy when he insists that
agriculture is not a form of industry. Land should *not* be seen as fixed
capital or even as a factor of production at all; instead, land is the basis
for life itself, a precondition for productive economic life, and not
merely one among many factors available for productive appropriation.
Diesing believes the agrarian component of Schumacher's thought is its
essential philosophical theme. The more celebrated work on appropriate technology originates from Schumacher's perspective of agriculture,
rather than the reverse.

Neither Schumacher nor Leopold was professionally trained in philosophy. The work of translating their insights into systematic ethical
theories has been undertaken by a score or more of philosophers who
have produced an enormous collection of work, mostly in the past
decade. One of the main bones of philosophical contention has been
whether the kinds of environmental goals espoused by Leopold and
Schumacher can be adequately justified by any application of anthropocentric thinking, utilitarian or otherwise. The first extended philosophical treatment of environmental ethics focused on precisely this
issue. In *Man's Responsibility for Nature* (1974) Australian philosopher
John Passmore responded to the growing tendency of environmental
scientists and activists to believe that general acceptance of a new, non-
Western, and non-anthropocentric ethic was necessary to solve a looming
environmental crisis. Passmore responded to Lynn White Jr.'s claim in
an often cited article (1967) that the Judeo-Christian tradition was ultimately responsible for environmental degradation because that tradition's anthropocentrism encouraged a despotic attitude toward nature.

Passmore argued that a culture cannot decide to adopt a new world
view the way a person decides to put on a new coat and that any new
ethic must grow from indigenous Western "seeds." He emphasized,
however, that Western traditions are more diverse than White had
indicated; in particular, the Christian tradition includes a "seed" of
"stewardship," according to which human beings must safeguard natural resources for both (human) posterity and the glory of God. Such a
Western tradition could be used to justify sound environmental policy,
he argued, without abandoning anthropocentrism. Passmore concluded
that to the extent that the goals of the environmental movement are
justifiable, they can be justified within the Western tradition, and that if
a radically new ethic is required to justify them, then they are unjustifiable. In particular, he concluded that some of the more idealistic
goals of the environmental movement—things like setting aside large
tracts of undeveloped land and preserving every remaining species

from extinction—indeed are unjustifiable.

Since Passmore's work, the literature in environmental ethics has virtually exploded, and a lion's share of philosophers' attention has been devoted to the issues on which Passmore focused: the extent to which the goals of the environmental movement can be justified from within familiar Western traditions, and what theoretical innovations, if any, would be necessary to justify them. Early papers in this genre include Christopher D. Stone's "Should Trees Have Standing? Toward Legal Rights for Natural Objects" (1972); Arne Naess' "The Shallow and the Deep, Long-Range Ecology Movement: A Summary" (1973); Richard Routley's "Is There a Need for a New, and Environmental Ethic?" (1973); and Holmes Rolston's "Is There an Ecological Ethic?" (1975). Rolston and J. Baird Callicott have subsequently emerged as the most prominent anti-anthropocentrists. Rolston summarized his mature view in readable but philosophically imprecise form in *Environmental Ethics* (Rolston, 1988). Callicott's work is discussed at some length in the final section of this chapter.

Others have countered Passmore by arguing that the more ambitious goals of the environmental movement can be justified from within an anthropocentric stance. A splendid example is Bryan Norton's recent book, *Why Preserve Natural Variety?* (1987). Norton lucidly critiques each of the major non-anthropocentric approaches and then appeals to ecological science in defense of an aggressive, habitat-based approach to species preservation, an approach which envisions a global patchwork of developed and undeveloped areas reminiscent of Leopold's vision of "biotic farming."

While it is misleading to suggest that much theoretical consensus exists among environmental ethicists at this time, general (but not universal) agreement has emerged on one point. Because so many environmental ethicists have believed that developing a non-anthropocentric ethic would be a necessary step in solving the ecological crisis, some believed early on that an animal rights perspective would be helpful in environmental ethics. Today, however, the leading thinkers in both fields agree that animal rights and environmental ethics are based on fundamentally different ethical principles, although there is disagreement about whether those principles are inconsistent in their practical implications. Compare Callicott (1980) and Regan (1983) (in section 9.3) with Sapontzis (1987) (in chapter 14) and Stone (1987). These arguments illustrate that animal rights-based criticisms of agricultural practices and research require separate treatment.

Readers who want a more complete survey of environmental ethics literature should consult introductory essays by Callicott (1986), Hargrove (1987), and Sagoff (1988a; 1988b). Hargrove's *Foundations of Environ-*

mental Ethics (1989) is a book-length introduction, and Davis's *Ecophi-losophy* (1989) is a book-length annotated bibliography. Anthologies in environmental ethics reprint many of the most important arguments. (See Scherer & Attig [1983], and VanDeVeer & Pierce [1986]). Regan's *Earthbound* (1984) is a collection of essays written at an introductory level, each of which references an extensive amount of literature. Important works on responsibilities to future generations, a topic that has been omitted from this chapter, are anthologized in Partridge (1981).

Animal Welfare and Agricultural Research

It is becoming increasingly common to distinguish between "animal welfare advocates" and "animal rights advocates." However, the distinction usually is made in terms of the practical conclusions the two groups reach and in the ways by which they prosecute their political agendas, rather than in terms of their basic philosophical differences. The expression "animal rights advocates" refers to people who favor the total abolition of animal agriculture and experimentation and who are willing to "liberate" animals and destroy public property in pursuit of this goal. "Animal welfare advocates," in contrast, adopt a more moderate position, both in terms of their goals and the ways they pursue them.

Activists concerned with the treatment of non-human animals in research and agriculture, especially the more radical ones, are often dismissed as being anti-scientific and irrational, or at the mercy of their emotions. The activists often dismiss researchers as unfeeling moral monsters on the same plane as the Nazi doctors who performed experiments on Jews. Neither viewpoint is conducive to understanding how concern for animal interests might affect agriculture and agricultural research. What is needed instead is careful, open-minded examination of the philosophical views behind the clashing views of experimenters and animal activists.

Works on animal rights/welfare vary greatly in the extent to which they engage in philosophical analysis. Some books devote little or no space to the philosophical basis of the animal rights/welfare movement. They simply catalogue the woes animals experience in modern intensive agricultural systems, without explaining why animal suffering should matter or how heavily it should weigh in the moral evaluation of agricultural practices. Examples are Harrison (1964), Singer and Mason (1980), and a host of books by Michael W. Fox (1976; 1980; 1986). Fox, who is a veterinarian, should not be confused with Michael A. Fox, a Canadian philosopher who has written the definitive defense of animal experimentation from an anthropocentric view (M.A. Fox, 1986).

Peter Singer's *Animal Liberation* (1975; 1990) contains the classic statement of the philosophical rationale for the animal welfare position. Singer's book mixes philosophical analysis with a catalogue of woes. Less than half of the book is devoted to the philosophical basis of the animal welfare position. More than half is devoted to description of problems and suggestions for living a vegetarian lifestyle. However, Singer is a professional philosopher, and his philosophical analysis builds on an important tradition in Western ethical theory. This is why his book is so important as a defense of the animal welfare position.

In the most philosophically important chapter of the book, Singer argues that the capacity for suffering and enjoyment is a necessary and sufficient condition for having interests.

> It would be nonsense to say that it was not in the interests of stone to be kicked along the road by a schoolboy. A stone does not have interests because it cannot suffer. Nothing that we can do to it could possibly make any difference to its welfare. A mouse, on the other hand, does have an interest in not being kicked along the road, because it will suffer if it is (1975, p. 8).

If a being suffers, there no justification for failing to take its interests into account, or for not weighing its suffering equally with the similar suffering of a human being. Singer does not claim that animals suffer in the same ways humans would in similar situations (for example, that a pig suffers when deprived of an education). He claims only that, whatever its cause, a given amount of animal suffering should be weighed equally against the same amount of human suffering.

After defending these philosophical claims in the first chapter of the book, Singer devotes an entire chapter to vivid portrayals of the ways animals suffer during intensive rearing and slaughtering practices. His argument comes to this: If we are not justified in inflicting this much suffering on human beings for the sake of obtaining inexpensive meat, then we are not justified in inflicting this much suffering on animals for the same purpose. Singer concludes that only because "Most human beings are speciesists" (p. 9) (i.e., they discount animal suffering as compared to similar human suffering), we fail to see that most current animal agriculture, and most research aimed at increased efficiency, is unjustifiable.

Singer often writes as if he opposes all slaughter of higher animals (meaning at least all mammals and birds) for meat, but his philosophical principles do not adequately support such blanket opposition. A logically consistent application of Singer's ethical principles would permit the painless slaughter of humanely raised animals: A happy life and a painless death (however premature) introduces no evil into the world, the meat brings pleasure to human beings, and the slaughtered animal can be replaced with an equally happy animal. Similarly, Singer often writes as if he opposes all animal experimentation, but a consistent application of his principles would justify experimentation which promised to significantly reduce the suffering of human beings and/or animals.

Singer's philosophical principles provide reasons for criticizing modern, intensive animal agriculture and the research supporting it. They do not give him a basis for rejecting the practice of animal slaughter entirely, or for refusing to conduct well-designed research that promises to improve the condition of humans and/or animals. It is in this sense that Singer's philosophy supports the moderate view associated with animal welfare advocates. Taking animal suffering into consideration does not require the total abolition of animal agriculture and related research, but it does require a sincere and systematic attempt to mitigate the animal suffering involved.

The practical implications of the moderate view have been explored by philosopher Bernard Rollin (1981). Although Rollin employs a broader notion of interests than Singer, his view is based on balancing human interests against the interests of non-human animals, and his emphasis is on improving the conditions in which agricultural animals live. Relevant research on applicable agricultural technologies has been conducted by animal scientists such as Theodore Friend and Stanley Curtis, though they have not articulated a philosophical rationale for their work.

Animal Rights and Agricultural Research

The most complete statement of the philosophical position that supports the stronger conclusions of "animal rights activists" is found in Tom Regan's book, *The Case for Animal Rights* (1983). Like Aldo Leopold and in sharp contrast to Singer, Regan appeals to philosophical arguments that reject utilitarianism. Singer's view is utilitarian in so far as he stresses minimizing the pain or suffering caused by intensive farming methods. Regan rejects the utilitarians' definition of right actions as those which maximize aggregate utility, however, and this rejection of utilitarianism (coupled with a broader view of what constitutes harm) leads Regan to conclude that all commercial animal agriculture and almost all animal research must be abolished.

Like many philosophers who have written about the moral rights of humans, Regan rejects utilitarianism because, if taken to its logical conclusion, it would leave no room for the important ethical notions of justice and rights. For instance, if the greater good could be realized by convicting and punishing an innocent man, or by enslaving a portion of the population, then a consistent utilitarian would have to agree that such acts are morally justified. Many philosophers have based their rejection of utilitarianism on the theory's capacity to rationalize such unpalatable results. When Regan applies what he calls "the rights view" to the issue of animal agriculture, he is following in the footsteps of a longstanding tradition of philosophical authors.

Regan believes that whenever harm must be inflicted on an individual, respecting the individual's moral rights entails acting in accordance with two moral principles, one of which is particularly relevant to animal agriculture: the "worse-off principle." The harm humans would suffer by giving up nutritionally unnecessary meat products is not comparable to the harm animals suffer when slaughtered, particularly when harm is conceived, not merely as pain that is felt, but as a restriction in the range of desires that may be satisfied during the course of a life.

Regan's conception of harm differs from Singer's in that death is the greatest harm an animal (or human) can suffer, whether that death is painless or not. The "worse-off principle" states that respecting an individual's moral rights requires that we desist from acts which harm that individual in ways which are non-comparably worse than the harm done to any individual under any other alternative. Since the harm a mammal or a bird suffers when it is killed prematurely is non-comparably worse than the harm any human being would suffer in giving up meat, the worse-off principle requires the total abolition of animal slaughter, and thus also the total abolition of research aimed at increasing the efficiency of industries based on animal slaughter.

It is Regan's "rights view," rather than Singer's utilitarian ethic of welfare, that provides a principled defense of the conclusions identified with "animal rights advocates." When such a position is defended, a host of objections invariably arise in the minds of skeptical listeners. While some of these objections have merit, many are based on an inadequate understanding of the philosophical principles behind the view. An objection that comes to mind, for example, is that farm animals must, like wild game, be culled to prevent ecologically detrimental overpopulation. But this objection has no force against Regan's position, once his philosophical principles are understood. Overpopulation could be prevented through sterilization, and although sterilization constitutes a harm (since it restricts the range of desire satisfaction available to the animal or individual), Regan's worse-off principle would permit sterilization of individuals when it was necessary to prevent non-comparably worse harm to individuals in an overpopulated herd.

A more serious objection is that in reaching his conclusions, Regan has ignored harms to human beings, in particular those most dependent, economically, on animal agriculture. The final chapter of Regan's book contains explicit and detailed responses to such objections. These cannot be reviewed here. We refer to them only to emphasize that the strong conclusions reached by "animal rights advocates" cannot be dismissed as irrational, purely emotional, or unthinking. As represented in Regan's book, the animal rights position is a rational, coherent view, based on an important tradition in ethical philosophy.

Environmental Ethics and Modern Agriculture

As noted previously, most environmental critics of agriculture have taken an approach to ethics that assumes the priority of human interests, and adopts the utilitarian method of evaluating an act in terms of its expected outcomes. This approach makes the environmentalist's argument against productivity different from other criticisms discussed in this book. Advocates of small farms and Third World peasants often stress justice and duties of respect, two concepts that always have seemed at odds with the utilitarian emphasis on optimizing outcomes. In citing externalities, environmentalists are making charges against agriculture that are conceptually more congenial to the utilitarian way of evaluating choices, however bitter they may have been to accept. The approach that environmental complaints are at least conceptually compatible with the traditional pattern of thinking in the USDA/land-grant system may explain why these complaints have been more readily taken to heart by agricultural scientists and agricultural administrators.

As we have observed, however, a substantial number of authors in animal welfare/rights and environmental ethics do not accept the utilitarian approach to ethics and base their complaints against agriculture on very different foundations. Relatively few of the authors who adopt anti-utilitarian philosophies of ethics have written explicitly about agriculture. To be sure, Peter Singer includes extensive sections on agriculture in *Animal Liberation*, and has, with Jim Mason (1980), written an entire book on animal agriculture. By extending the utilitarian calculus to include impacts upon non-humans, Singer identifies a new class of externalities, but in accepting the utilitarian emphasis upon optimizing benefits, he has adopted an ethical framework that is consistent with the standard pattern of the USDA/land-grant way of thinking. His message is one of including costs to non-human animals in the optimizing calculation. As such, his message not a criticism of how agricultural scientists have set priorities, but a plea (much like those of

Rachel Carson or David Pimentel) for scientists to do a better job of including all the costs when they make their assessments.

Because the total rejection of utilitarian thinking represents the most serious challenge to traditional agricultural research policy, it is important to conclude this chapter with a more detailed look at criticisms that go beyond the citation of externalities. Here we will examine sources in which authors have made specified complaints against agriculture and agricultural research, rather than broad claims from which implications can be drawn. This literature rapidly narrows down to those books and articles which have advanced ecocentric ethical arguments against the goal of increasing productivity. Although Regan has written a piece applying his animal rights view specifically to agriculture, the animal rights position leaves no room for reforming research goals in animal agriculture; animal agriculture must simply be abandoned. The balance of the chapter, thus, focuses exclusively on environmental ethics literature.

Given the enormous significance that agriculture and environment have for one another, the relative paucity of philosophical work on the relationship is surprising. Even journals such as *Environmental Ethics* and *Agriculture and Human Values* devote an overwhelming majority of entries to other topics. Aside from articles that fit more comfortably in the "externalities" tradition, *Agriculture and Human Values* has published articles by Ebenreck (1984), Blatz (1984), Meeks (1985), Worster (1985), Healy (1985), and Thompson (1986). Although agricultural ethics has addressed the question of sustainable agriculture, it has tended to do so from the perspective of internalizing long-term production externalities (Burkhardt, 1989).

Among those who have argued eloquently that environmental ethics is *not* simply an extension of traditional ethics is J. Baird Callicott. This theme emerged clearly in a 1980 paper entitled "Animal Liberation: A Triangular Affair," in which Callicott criticized the animal rights/animal welfare movement. Callicott faulted both themes in the animal welfare/ rights literature, noting that neither could account for the concern that game management professionals have for wildlife populations and endangered species. Callicott argued that an overriding moral concern for the rights of individual animals would preclude many of the game management strategies for culling populations that were exceeding carrying capacity and for protecting species. Conservationists, in other words, are concerned about the role played by species in the balance of nature, and their concern (if any) for individual animals is purely a consequence of their relation to the whole (Callicott, 1980).

Like many philosophers who have written on the subject, Callicott's original interests were in wildlife and conservation, but, unlike others, he has written at some length on agriculture. He was, in part, led there by Leopold. Leopold's career as a conservationist was based upon his pioneering studies in game management. One of Leopold's early discoveries was that intensive game management was the primary culprit in the precipitous decline of predator species. It was not that predators were "bad citizens," but rather that the context in which predator and prey had always interacted with one another was being dramatically upset (Callicott, 1987, p. 286). Callicott cites this theme in Leopold's work as evidence that Leopold should not be regarded as simply extending the scope of moral evaluation, but as proposing an alternative that stresses a holistic, community-based approach to value.

In his recent articles on agriculture, Callicott argues that we must resist the intellectual temptations of reductionism. In the sciences, reductionism leads to a view of nature which is purely mechanical and stripped of value that might be apparent, or even obvious, when natural phenomena are considered in the context of daily life. In ethical theory, reductionism leads to an inappropriate emphasis on the moral units of pleasure, satisfaction, and rights. In tandem, scientific and ethical reductionism combine to produce a point of view that is oblivious to the value of preserving the whole (Callicott, 1987; 1988).

Callicott's most recent work is particularly relevant for research policy. His 1990 article, "The Metaphysical Transition in Farming: From the Newtonian-Mechanical to the Eltonian Ecological," documents two perspectives on biological process. Callicott believes that agricultural scientists have persisted in applying the "intellectual template" of classical Newtonian physics long after its demise in physical theory. According to the Newtonian world view, knowledge of the natural world is exhaustively achieved when scientists obtain an account of the mechanisms that underlie natural events. The alternative view, which Callicott names after Charles Elton, notes that emphasis upon mechanisms will never lead one to recognize processes and interrelationships that occur at the level of whole populations, or ecological systems. Following a line of thought originally voiced by Levins and Lewontin (1985) and by Wes Jackson (1980; 1987), Callicott argues that reform in agricultural science requires a rejection of the Newtonian view of natural process.

It is important to note that Callicott's analysis deviates from criticisms that identify modern agriculture's problems with its emphasis upon productivity. Although Callicott is not slow to indict those who emphasize yield-increasing research and the pursuit of profit, (see especially Callicott, 1988; and Lappe and Callicott, 1989), the real problem is not the agricultural scientist's emphasis upon farm productivity. Rather, it is a view of nature that the agricultural researcher shares with many mainline biologists. Callicott implies that if researchers would adopt scientific views that are more commensurate with the evidence both at a micro level (in theoretical physics) and at a macro level (in ecology), they would shift the focus of their research to more environmentally benign and sustainable strategies. It is a close wedding of beliefs about nature (metaphysics) and beliefs about the good (ethics) that will produce this change, rather than an alteration of ethical values, as understood in the conventional sense.

Conclusion

In concluding this review of literature in animal welfare and environmental ethics, it is important to note a connection with the literature on agrarianism surveyed in this volume by Nancy Shankle. Wendell Berry, whose works are discussed at some length in Shankle's chapter, has been classified as an environmentalist by others (McKibben, 1990). Berry's writings about small farms and farm communities emphasize the way that traditional farming defines a role for each member of the family. Individual role responsibility is defined in relation to the unified needs of the family as a unit. The family's survival, in turn, depends on its interaction with the natural environment. Husbands, wives, sons, and daughters of the farmstead must see a close connection between their contributions to the family livelihood and the cycles and requirements of the natural world. In this way, agrarian virtues speak to the issue of responsibilities for nature.

In point of fact, Callicott's criticism of reductionism in agricultural science can also be linked to Berry's agrarian holism. The connection requires emphasis on the problems caused by a more specialized division of labor. Industrialization of agriculture destroys the integration between the farmstead's activity and the natural world. Farms become production enterprises aimed at producing profit. This fact separates or divides the roles required for operating the farm (manager, laborer, technical expert) from the roles that define family relationships (breadwinner, parent, child). In addition to becoming divided from one another, farm and family roles become increasingly specialized in relation to their internal function, and individuals have difficulty understanding their personal responsibility for the whole enterprise. Thus, the breadwinner specializes in family survival (now ensured through cash income), and the child is deprived of role activities that would promote the development of a broader moral and environmental sense of duty. The manager concentrates on the balance sheets, and thus may fail to associate that activity with the farm's dependence upon nature's abundance (see Berry 1977; Burkhardt, this volume; and Shankle, this volume).

Callicott's criticism of reductionism in science is an additional species of the same observation. Scientists who become more specialized also become more prone to errors of omission. They may overlook connections and context that, in a simpler age, made the moral significance of the whole obvious to anyone. The connection between scientific reductionism and division of labor on the farm and within the family makes each theme appear as only one example in a systematic critique of the modern age. Here, the rejection of utilitarian thinking is crucial, because utilitarian thinking portrays moral responsibility in terms of separable goods that may be pursued in specialized ways. This broader argument has been intellectually advanced in books such as *Habits of the Heart* (Bellah et al., 1985) and *After Virtue* (MacIntyre, 1981).

These final themes are deeply subtle and comprehensive. While we applaud readers of this chapter who pursue the issues of agricultural research planning to this level of philosophical depth, we must note that such themes are both difficult and controversial in themselves. This does not mean that they are unimportant for agricultural research; they may be the most important of all. The broad cultural critique that emerges when agrarian and environmental holism are conjoined goes far beyond the constraints of this chapter. However, with this passing note on the possibility of such a conjunction, the discussion must end.

11

U.S. Agricultural Research Policy and International Distributive Justice

G. Lain Ellis

Hearken and hear then, said [Thrasymachus]. I affirm that the just is nothing else than the advantage of the stronger. Well, why don't you applaud? Nay, you'll do anything but that.

Plato, Republic

Thrasymachus apparently believed that his audience would not applaud because to do so would require acknowledging a universal truth too painful for well-meaning people to bear. The reader who agrees can stop here. Still, there may be people who believe that justice is not the same thing as power, and that seeking the former (however imperfectly it may be achieved in practice) is too important to allow the latter to stand in the way. This essay is intended to provide the agricultural scientist with an introductory inventory of issues and sources relevant to the relationship between U.S. agricultural research policy and distributive justice in the Third World. One object of the essay is to describe justice-related perspectives that have been expressed in the research policy, agricultural development, and philosophical literatures. These bodies of literature, however, are individually massive, with the central issues in one often discussed only in passing (if at all) in the others. Consequently, another object is to identify issues and values that connect and contrast the perspectives described here. Because the intent of this essay is to be informative, I have attempted to avoid arguing for or against any given position. But, as I am no friend of

Thrasymachus and have definite beliefs about justice, the attempt at neutrality has been incompletely successful.

The essay is structured in three sections. The first is an extended discussion of V.W. Ruttan's *Agricultural Research Policy and Development*, a work that connects agricultural research policy for Third World contexts with a view of moral responsibility that centers around research entrepreneurship as a means for meeting present and future needs. This section identifies some of the parallels between Ruttan's position on research policy for the Third World and the values discussed in the introduction of this volume. Next, a review of K.A. Dahlberg's *Beyond the Green Revolution* introduces controversies emerging from the Green Revolution. This discussion is followed by accounts of agricultural development based on high-yield variety (HYV) packages and free markets, and critiques of that approach. This section identifies the sets of empirical variables that each of the two approaches regards as internal to research problem formulation and to the evaluation of research. The third section examines some alternative concepts relevant to just development and agricultural research policy. It focuses on how the utilitarian and research entrepreneurship values of U.S. agricultural research policy can be understood in terms of the critical literature.

The essay centers around descriptions of a relatively small number of books and articles. The descriptions vary considerably in detail because of a need to establish a representative conceptual basis for comparison, and the works covered in detail were selected for their historical and/or conceptual relevance. This format was chosen because the issues are complex, and a long list of references without much description would be neither particularly informative nor interesting. The essay also includes extensive quotes to give the reader a flavor for the works being described. Additional references have been included to direct the reader to related sources. With some luck, the combination of described and mentioned works will deepen the reader's appreciation of the conceptual foundations that underlie different evaluations of U.S. research policy as it pertains to distributive justice.

Agricultural Research Policy for Third World Contexts

A major work on agricultural research policy for developing nations is Vernon W. Ruttan's *Agricultural Research Policy and Development* (1987), which serves here as the point of departure for discussing issues related to distributive justice in the Third World. This point of departure was chosen for several reasons. Ruttan has been and continues to be a major figure in the literature on research policy for both the U.S. and the Third World. Ruttan (1987) articulates and defends a detailed argument for Third World research policy. This argument draws extensively on the widely perceived successes of the USDA/land-grant system, and may be construed as reflecting an application of U.S. research policy to Third World contexts. Finally, Ruttan (1987) offers an explicit argument for a moral position that justifies this application.

Ruttan's study begins with an overview of the development of national research systems, from which he derives four models for the organization of agricultural research. Ruttan notes several areas of stress and conflict that are present in some degree in each of the models. The proper relationship between research, education, and extension is difficult to establish. Graduate education and research are often highly complementary activities for agricultural scientists, but many scientists find it difficult to link research with undergraduate education and extension activities. Second, stress emerges from the degree of centralization or decentralization in a given system because of difficulties in achieving a fit between national and regional priorities, and because increased bureaucratic restraints tend to accompany increased centralization. Third, a conflict is present in the allocation of research resources to the basic and applied research that emerges because the personal and professional goals of individual scientists (whose interests are often tied to advances in basic research; see Busch and Lacy, 1983; and Thompson, Ellis, and Stout, this volume) may diverge from objectives set by society.

These stresses lead Ruttan to express several concerns facing the development of national research systems in the Third World. He notes that there is a tendency to overinvest in facilities relative to available scientific staff because development agencies find it easier to invest in facilities than to support the development of human capital or research programs. (See de Janvry and Dethier [1985] for other aspects of underinvestment in Third World research systems.) As a result, the reduction of assistance may be accompanied by erosion of research capacity. Development, maintenance, and implementation of research capacity is further hindered because agricultural education, research, extension, and technology transfer functions are often poorly articulated. Moreover, research facilities often may be established without considering whether their locations have appropriate human and physical infrastructural characteristics, or whether their soil/water/climate regimes are representative of the commodity and/or agricultural resource problems that have been identified. Significant problems also exist in establishing research priorities, and budgets often are out of step with the importance of the problems to be solved. Typically, there is no mechanism to adjust research priorities and capacity to changing resource endowments, or to articulate public and private sector research efforts. Furthermore, routine research activities and research entrepreneurship are often stifled by administrative burdens.

Ruttan (1987:52) suggests that these concerns

> reflect a lack of systematic analysis and planning in the design and management of research systems....In the research planning staffs that have successfully struggled with the research resource allocation problem, it has become increasingly obvious that effective research planning requires close collaboration between natural and social scientists and among agronomists, engineers and planners.

He claims (1987:52) the research planning problem stems from a failure to distinguish between two basic questions in the evaluation of research, especially in relation to the persons best-placed to answer them:

> One question is what are the possibilities of advancing knowledge or technology if resources are allocated to a particular commodity problem or discipline? Such questions can only be answered [effectively] by scientists who are on the leading edge of the research discipline or problem being considered. The intuitive judgments of research administrators and planners are rarely adequate to answer [them].

> A second question is what will be the value to society of the new knowledge or the new technology if the research effort is successful? The intuitive insights of research scientists and administrators are no more reliable in answering questions of value than are the intuitive insights of research planners in evaluating scientific or technical potential. Many of the arguments about research resource allocation flounder on the failure of the participants to clearly recognize the distinction between these two questions and the differences in expertise and judgment that must be brought to bear in responding to them.

These questions reflect a tension between the goals of agricultural science and society. Management and planning have become increasingly centralized since the 1960s. Ruttan claims that returns from decentralized research planning have been high, and that increasing the allocation of resources for centralized management and planning in turn requires demonstrating that these resources return greater benefits than when they are devoted directly to research.

Ruttan believes that when they evaluate the impact of potential research, scientists and administrators are most comfortable answering the first question—the potential contribution to knowledge or technology. Scientists on the forefront of research are well-placed to evaluate the potential for further advances in knowledge, and there are methods to evaluate the rate, direction, and social distribution of productivity growth from research. (See Schuh and Tollini [1979], de Janvry and Dethier [1985], Madden [1986], and Dalrymple [1977] for discussions of evaluating returns on research.) On the other hand, no good methods are available for evaluating impacts with respect to some kinds of social goals (e.g., aesthetic and ethical impacts), especially when those goals have not been articulated in the political processes that establish research priorities. Thus, research planners who must allocate resources with an eye toward the second question often do not have sufficient or appropriate information.

However, as resource-allocation evaluation methods become more prevalent, Ruttan warns (1987:106) that they cannot become mechanical devices for making planning decisions:

> Judgments about the priority and the value of public-sector agricultural research, relative to other demands for public resources, must, in most countries, come out of an intricate bargaining process that goes on between national (or state or provincial) legislative bodies and executive agencies. Judgments about the relative emphasis that should be given to saving labor, saving land and saving energy—as well as...to the effects of production technologies on the health and welfare of producers and consumers, and the impact of agricultural technology on environmental amenities and on the structure of rural communities—must also come out of this same bargaining process....The political dialogue leading to research resource allocation should be fully informed about the costs and benefits of research resource allocation decisions. Information on the historical and potential impact of technical change can represent a valuable input into these bargaining processes, even if it serves primarily to keep the debate reasonably honest.

Mobilization of political support is a key element in Ruttan's vision of resource allocation and moral responsibility in agricultural research (see also Ruttan, this volume). One of the tasks facing the research institution is to build political support for research. He notes that there is an infinite range of scientifically interesting agricultural problems, but that not all of these problems are equally important. Ruttan claims (1987:138-139) that if the political process identifies major social problems that can be addressed by agricultural research, then "if scientists can convince people that research can [solve those problems] there should be political support for money to finance research. Most morally responsible scientists would have no difficulty working on these problems." Indeed, Ruttan claims (1987:221), an "agricultural research system should not, and cannot, be designed without taking into consideration the needs and capacities of the broader society of which it is a part."

Ruttan also acknowledges that political influence on agricultural research is not evenly distributed. If administrators do not allocate research activities to problems of well-organized interest groups, they may risk reduction of their overall budget. However, Ruttan further claims (1987:139) that if they allocate too much research toward narrow interests, administrators are not "exercising [their] moral responsibility on major problems facing the general public and will lose [its] support, which will also lead to a reduced budget." Research administrators must balance the agenda to support research for a wide range of social goals. Balancing the agenda requires administrators to support budget requests with information regarding the social value and impact of proposed research, which in turn requires access to multidisciplinary research (including social science) that demonstrates impacts, including prospects for the adoption of research results. For example, a large, socially mandated budget for small-farm research will accrue little return if society systematically eliminates small farms because of its land-tenure, credit, and other policies. Thus, the political role of the administrator involves determining what society wants and informing society of the likelihood that goals can be met by agricultural research.

Not only must administrators communicate with external constituencies; they also must communicate those constituencies' goals to individual scientists. However, Ruttan does not believe that administrators should dictate the research agenda in detail. Such administrative control is likely to stifle the research entrepreneurship that serves as the basis for scientific and technical advances. In Ruttan's view, the high rates of return from decentralized research warrant having moral responsibility for research "rest lightly" on scientists and administrators. This is not to say that the agenda can remain detached from ethical and social issues; rather, the research community must become increasingly sensitive to these issues in order for scientists to identify research solutions consistent with social goals. However, Ruttan believes that research entrepreneurship is too important to allow a heavy burden of moral responsibility to constrain the creativity of individual scientists. He suggests instead that the impact of research can best be controlled at the adoption stage.

Ruttan's position reflects a number of elements of the structure of U.S. research policy as described in the introduction of this volume. A central premise of the argument is his description of the administrator's position in establishing and implementing the research agenda. According to Ruttan, the social goals for research should be established in an open political debate. A research administrator is uniquely positioned to gather a broad range of information about the options that can be applied toward social goals, and the administrator's primary role is to ensure that the debate is well informed about the prospects for meeting those goals through agricultural research. Once goals are set, administration must be sufficiently centralized to coordinate research to reflect social goals, but also sufficiently decentralized to promote research entrepreneurship as a source of optimal social benefits.

Thus, Ruttan's position is based on the premises that (1) the members of society are best-placed to determine what the social goals for research are; (2) administrators are best-placed to inform members of society about the potential for serving those goals through research, and to communicate those goals to scientists; and (3) individual scientists are best-placed to identify specific solutions for research goals. A crucial issue for U.S. research policy and Third World justice, therefore, is whether the values of utilitarianism and research entrepreneurship provide a sufficient or appropriate foundation for research for development. Before we can turn to this issue, it is necessary to examine some of the literature on Third World agricultural development.

Exporting Science, Technology, and Values

Beyond the Green Revolution (1979), by Kenneth Dahlberg, is a critique of the technology development and transfer orientation that has been the main emphasis of agricultural development in poor countries. Dahlberg believes that the history of development represents an attempt to graft Western practices onto Third World contexts without sufficient understanding of the values involved and the long-term prospects that result.

Dahlberg begins with an "ecology of theories," in which he claims that belief in the universality of Western science is itself an outgrowth of other Western cultural tendencies, especially the notions of "progress" and dominion over the earth. He claims that a belief in the superiority of Western society, buttressed by its historical successes in colonization and economic development, contributes to a belief in the superiority of science over traditional, non-Western or non-scientific knowledge systems. Moreover, increasing specialization within Western societies has affected the structure and content of science through the infusion of reductionism and disciplinary specialization into the research community. Scientific research institutions, especially when there is an interdependence among theory, complex technologies, and bureaucracies, have "a variety of vested interests that make the introduction of new scientific theories difficult" (Dahlberg, 1979:8-9). The network of vested interests is further reinforced by another Western institution—the nation-state—that was diffused to the non-Western world through colonialism and subsequent decolonization. The nation-state brings with it another set of vested interests linked historically to Western training of non-Western leaders.

Despite massive efforts to develop non-Western nations, the international economy has remained Western both in structure and in delivery of most benefits to Western industrial nations. Dahlberg believes that the economic theory underlying many development efforts must be rethought for two closely linked reasons. The rise of the free-trade doctrines at the heart of economic theory was associated historically with the commercial and industrial success of countries that were already in strong trading positions, suggesting that free-trade values express the vested interests of the Western nations that implemented them rather than universal principles. Furthermore, the free-trade model is based on the Industrial Revolution, and ignores the earlier agricultural and commercial revolutions that made it possible. Indeed, as the Industrial Revolution progressed, agriculture was reduced to a "Cinderella-like position— an indispensable but definitely lower-status stepsister" (Dahlberg, 1979:11).

Thus, Dahlberg sees the growth of Western economic theory and disciplinary agricultural science as the result of complex cultural and historical trends that have been mutually reinforcing. He believes that the failure to achieve significant development success is a reflection both of the resistance of vested interests (including scientific interests), and of the non-universality of the conceptual structures being used to develop agriculture in poor countries. As a case in point, Dahlberg examines the Rockefeller Foundation's project to increase wheat production in Mexico in the 1940s. The program was successful in developing high-yield varieties (HYVs) that increased Mexican wheat production. However, an unforeseen (or foreseen, but ignored; see DeWalt, this volume) outcome of the project was that wealthier farmers on better lands were able to take advantage of the HYVs, whereas poorer farmers on poorer lands were not. This outcome, Dahlberg claims, was a direct result of the "built-in" assumptions and values of Western science, assumptions and values that reflect an emphasis on productivity, the distance between researchers and their clientele, and the constraints of specialized disciplinary research.

Most importantly, however, Dahlberg believes that the Green Revolution's emphasis on transferring HYV packages is accompanied by an inadequate understanding of the unequal distribution of resources and benefits among local elites and poor people, and an inadequate comparison with the benefits that could be gained from applying the same resources to alternative plans (see also Kloppenburg, 1988a; in an interesting twist, Lele [1984] suggests that the industrial modernization favored by many African governments is an impediment to the implementation of rural development programs favored by some Western donor agencies.)

Dahlberg proposes a strategy of "contextual" analysis for Third World development. Contextual analysis recognizes the operation of an array of physical, biological, and cultural processes at a series of spatial and temporal scales. Solving any area's agricultural problems requires identifying the specific array of influences that have led to its current situation, and identifying short- and medium-range policies that are consistent with long-term development goals. Crucial to policy development, however, is a transformation of the conceptual base on which policy is formulated. Dahlberg believes that development must be based on sustainable agriculture by large numbers of rural people: If it has been difficult to achieve distributive justice by transferring industrial agriculture to the Third World, it has been even more difficult to transfer industrial society to the Third World to absorb rural people displaced by Green Revolution technologies. He asks, (1979:172-173),

> given the fact that information is power, how can the gathering and organization of data be structured so that it does not simply reinforce current maldistributions in agriculture...? The tendency for Western ...scientists to see data as "neutral" and abstract makes it difficult to face up to this question....[Any] serious approach to the question must necessarily involve the building up of new institutions, networks, and skills that will give the poor countries an independent capacity to assess and to influence the use to which such data are put. Such a data capacity must be independent in two ways: independent of dominant Western industrial conceptions and interests, and independent of dominant local elites.

Independent data systems, then, serve as the basis for empowering rural people in their own development process (see also George, 1984). Their own existing knowledge can be supplemented by relevant knowledge from the sciences to provide solutions adapted to local social, environmental, and economic needs and priorities. This includes being able to identify appropriate scales of production in contexts where Western industrial approaches to yield and efficiency may or may not be appropriate because the problems to be solved may or may not resemble the problems traditionally addressed by Western agricultural science. Furthermore, it includes achievement of an appropriate level of diversity and centralization in a process which recognizes that overall development is contingent on developing a broad subsistence base, in much the same way that Western development grew out of agrarian foundations. Thus, according to Dahlberg, one of the chief tasks of agenda-setting for the Third World is to ensure that research is appropriate to the context in which it is to be applied. This requires an examination of the assumptions and values that underlie research for development, especially in relation to the political, economic, and social impact that new technologies may have on their intended beneficiaries.

The Laissez Faire Approach to Development

An early study of the assumptions underlying development was Theodore W. Schultz's *Transforming Traditional Agriculture* (1964). In the early 1960s, many believed that traditional agriculture was an inefficient means for allocating resources, and that the primary obstacle to increased agricultural production by traditional farmers was their economic backwardness, which allegedly led to inefficient allocation of the factors of production at their disposal. Schultz defines traditional agriculture as a type of economic equilibrium in which farmers have used the productive factors available to them for generations without significant alteration. Because no new factors are introduced, the body of agricultural knowledge (including risks and uncertainties) is well-known and adjusted to available capital. Likewise, the set of preferences and motives for acquiring new sources of income are adapted to the available factors. In reaching equilibrium, the rate of return on investment in additional factors declines and reduces incentives to save for investment.

If this thesis is correct, it implies that traditional agriculture, contrary to widespread belief, is very efficient at allocating resources, a conclusion Schultz reaches on the basis of two case studies. Given the efficiency of traditional agriculture, Schultz proposes that to increase production, it is necessary to introduce new agricultural factors with a rate of return on investment high enough to warrant adoption. One such investment cost is initial capital. Since poor farmers (by definition) do not have it, and since increased investment in traditional agriculture does not provide enough return to raise it, initial capital must come from outside sources. Schultz claims that once investment becomes profitable, capital will be reproduced locally through agricultural growth.

Another investment cost is the generation of new factors such as biological technologies. Schultz regards farmers as "demanders" of new factors. But, he believes that new factors cannot be generated by private firms because the high costs of initial development are compounded by the (often high) costs of adapting new factors to local conditions. Much of the return on research to generate new factors, however, is captured by other firms and consumers. Moreover, the potential market may be too small to make distribution profitable: large numbers of poor farmers on very small farms need not add up to a large market for new factors. Therefore, Schultz believes, the development of new factors belongs properly to the public sector. Institutions to develop new factors may be established by foreign agencies, but in the long run, the poor countries must develop their own. After a sufficient level of development has been achieved, collaborative arrangements between public and private organizations may be necessary to continue the process of development and distribution of new factors.

Not only must the rate of return on new factors be high; it must be high enough to warrant the new risks and uncertainties that accompany them. According to Schultz, a principal obstacle to overcoming risk and uncertainty is the low level of education prevalent in many poor areas. Thus, another crucial investment is education, which Schultz views as the development of human capital. In addition to educating farmers to take advantage of new factors, Schultz sees two other educational needs. An effective extension service is needed to convey information to farmers, and poor countries must develop the capacity to train their own research and extension personnel because foreign educations may not provide the background needed to address the problems peculiar to a given developing country (see also Swanson, 1986; Esmay, this volume). Clearly, the costs of education are too high for the private sector to bear.

If Schultz rejects the Western stereotype of traditional agriculture, he nonetheless retains a Western emphasis on free markets as the appropriate mechanism for development. (See Ishikawa [1967] for a discussion of this aspect of Schultz's perspective.) Charles B. Shuman expresses the free-market orientation in "Food Aid and the Free Market" (1970). Like Schultz, Shuman believes that the primary obstacle to agricultural development and the alleviation of hunger is the lack of incentives for investment by poor farmers compounded by the lack of capital and/or credit for investment. Shuman claims that any amount of technological or other assistance is ultimately futile if it is not accompanied by free markets to provide incentives for poor farmers to take advantage of assistance. He argues that direct food aid retards investment in domestic production because by bypassing the market, it fails to translate the needs of poor people into effective demand that would stimulate local production. Direct food aid also isolates Third World governments from responsibility for the failures of their policies, and contributes to ongoing irresponsibility (see, e.g., Eicher, 1984). If aid is to be given, Shuman believes it should be in the form of direct cash grants to poor people who can then enter the food market and stimulate production by increasing the effective demand for food. (This recognizes the phenomenon, noted by Sen [1981], that adequate overall food supplies cannot avert famine if people are too poor to command a portion of the food supply.)

Intervention by Third World governments in their own markets also depresses the incentive to produce by keeping price levels too low to warrant the risk of investment. Shuman claims that interventionist policies further retard investment in agriculture because farmers will not adopt productive technologies if it is probable that future policy changes will negate the benefits to be gained from investment (as when land reform reduces holding size below the threshold of profitability for a given technology with given economy-of-scale characteristics). Furthermore, private interests do much of the applied research that adapts basic research to local contexts, and the absence of free markets provides disincentives for private international participation in development. Says Shuman (1970:151), "These...corporations and cooperatives that are so essential for a productive agriculture would move quickly to help develop the food-producing capacity of other nations if they could see any opportunity for a free market to come into being." Moreover, Shuman claims, agricultural development without development of other sectors (including free markets in those sectors) will perpetuate hunger because it will not provide opportunities for poor people to increase their effective demand by participating in the economy.

Supporters of the Green Revolution note that the adoption of HYVs in Third World contexts has resulted in greater yields. In "Accelerating Agricultural Research and Production in the Third World" (1986:9), Norman Borlaug, perhaps the most famous green revolutionary, says that "the greatest satisfaction of my scientific career has been to see India and Pakistan utilize [HYV packages to]...become self-sufficient in cereal production." He claims that in order for success to continue, it is necessary to make a long-term scientific commitment to recipient nations because programs may take a minimum of 10-15 years to achieve results. However, many Western university scientists who participate in research in Third World contexts are on leave for only a year or two, and many scientists in the International Agricultural Research Centers (IARCs) are shifted from assignment to assignment every two years or so.

This results in serious discontinuities in research efforts that often are aggravated by inadequate governmental support for national research systems, which, Borlaug says (1986:11), "is especially critical in the developing countries [because they] rarely have well-developed private sector agribusiness organizations to help in the task of agricultural development." Furthermore, Borlaug believes that a major role for the U.S. is to develop a larger base of Third World scientists. But, he suggests (1986:13) that Western educations often leave them ill-prepared to face research problems at home because they are over-specialized and lack the training necessary to perform as "integrators" of multidisciplinary research:

> Let's shift the focus away from the use of fancy equipment towards the development of intellectual skills that will allow these individuals 'to integrate research information more effectively across disciplines and thus help their countries achieve greater production impacts....[In] this increasingly interdependent world,...international trade is essential to the well-being of every nation, and history has proven that impoverished, chronically food-deficit countries do not become dynamic trading partners.

Some defenders of the *laissez faire* approach believe that it involves significant human costs. Garrett Hardin, in "Living on a Lifeboat" (1978a) and "Carrying Capacity as an Ethical Concept" (1978b), equates nations with lifeboats, some of which are occupied by rich people, others by poor people. Hardin notes (1978b:251) that "We can never do merely one thing," and claims that food aid, in the absence of other forms of aid or development, promotes growth among populations that already have exceeded the carrying capacity of their environments, while it simultaneously insulates Third World governments from pressures that would promote domestic production and agricultural research. Thus, by focusing on a single variable—food aid—rich nations postpone inevitable population reductions until some future time when many more people will starve.

Hardin claims that the rich nations apply a "triage" approach in which they decide which poor nations will be assisted, and which ones will not (see Paddock & Paddock, 1967). Hardin believes that food aid makes triage inevitable because people in the richer lifeboats are unlikely to sacrifice their standard of living to provide enough support for broad-based assistance that includes not only food, but also energy and support for the development of political, technological, and educational infrastructures. In the absence of such a commitment, "For posterity's sake we should never send food to any population that is beyond the reasonable carrying capacity of its land. The question of triage does not even arise" (Hardin, 1978b:256; see also Fletcher, 1977). Hardin believes that the development process is inevitably painful, but that a *laissez faire* approach is necessary to reduce the amount of long-run suffering. Hardin's position, therefore, is a utilitarian judgment based on the long-term consequences of present actions.

Thus, the *laissez faire* approach sees the development of domestic agricultural production as one of several simultaneously necessary elements in a development process conceptualized in terms of Western industrial and market models. This approach includes the notions that markets will serve poor people's food needs if they can exercise effective demand in the market, and that the nation-state is an appropriate unit of analysis to employ in development. It also includes the apparent assumptions that the goal of agricultural development is to increase the overall food supply, and that the way to do this is to make farming increasingly profitable through investment in technologies developed initially in a well-supported public sector, and subsequently adapted and diffused by private firms.

Distributive Impacts of Green Revolution Technologies

Discussion of the merits of Western agricultural technologies often revolves around inter- and intranational distributive issues. One of the claims typically raised against Green Revolution packages is that they intrinsically discriminate against members of the poorest sectors of Third World populations (see Grabowski, 1979). Yujiro Hayami, in his "Assessment of the Green Revolution" (1984), argues that adoption rates for HYV packages are independent of farm size, and that HYV packages are scale-neutral, i.e., they produce greater yields and profits on both large and small holdings. Furthermore, Hayami notes that while mechanization reduces labor for land preparation, demand for cultivation and harvest labor increases with the adoption of HYVs. (See Dorner [1983] for a discussion of scale-neutrality. Perelman [1977] claims that many farmers abandon HYV packages after a season or two because the increased costs and other demands are not offset by yields. Pearse [1980] disputes the increased demand for labor.)

Hayami claims that the principal source of growing inequality lies not in the HYV packages themselves, but rather stems from the land-saving nature of HYV packages amidst population growth. He suggests that the increased labor demand required by adoption of HYV packages has been insufficient to absorb population growth, which in turn has depressed the value of labor and negated the economic value of saving land. As a result, land ownership has become increasingly concentrated, and income increasingly skewed toward landlords who, lacking other readily accessible forms of investment to absorb their surpluses, buy more land and/or displace more tenants. Small farmers become increasingly concentrated on marginal holdings in which additional labor does not produce a sufficiently large return to sustain them on the land.

A. de Janvry and J.J. Dethier suggest that the innovation model (e.g., Ruttan & Hayami, 1972), on which defenses such as Hayami's are based, does not account for non-market influences on the adoption of new technologies. In *Technological Innovation in Agriculture* (1985; see also de Janvry, 1977), they note that poor Third World farmers constitute a large pool of latent demand for new technologies. However, market forces cannot translate latent demand into actual demand in which poor people lack sufficient access to resources, credit, indivisible technologies (e.g., tubewells), information, and political structures. Their lack of political strength often results in a lack of significant input into the research agenda-setting process, which may result in a research bias that favors more powerful groups with latent demand that is translated into actual demand for new technologies in the political process. As time passes, actualizing the latent demand of the poorest farmers may become more difficult. Return on investment in new technologies for subsistence farmers may be too low for too long (relative to returns from additional investment in already mature research streams) to be sustainable amidst competition for research resources in the agenda-setting process.

Andrew Pearse's *Seeds of Plenty, Seeds of Want* (1980; see also Hewitt de Alcantara, 1976) assesses the socio-economic constraints poor farmers face in the spread of Green Revolution technologies. Recalling a Biblical parable, Pearse refers to the polarization of richer and poorer farmers as the "talents effect": "For unto every one that hath shall be given and he shall have abundance: but from him that hath not shall be taken away even that which he hath" (Matthew, Ch. 25, v. 29). Pearse's study identifies factors that promote inequality when HYV packages are introduced into contexts already characterized by inequality.

Pearse argues (1980:181) that although HYV packages can increase yields and profits on large and small farms (as Hayami claims), this scale-neutrality is illusory because "the socio-economic magnitude of the cultivator is of the utmost importance for his economic success." He claims that adoption of HYV packages represents a radical discontinuity in the technologies, practices and information required by poor farmers. Not only is adoption accompanied by a set of unfamiliar risks, the packages themselves are often introduced in a "progressive farmer" strategy which targets likely innovators who then serve as models for less innovative farmers. However, Pearse claims that "progressive" farmers are more likely to have resources and skills that reduce risk rather than to have a pronounced predisposition for innovation. Thus, says Pearse, the strategy for introducing Green Revolution technologies initiates increasing polarization between relatively rich and poor farmers.

Furthermore, the "technological leap" required by HYV packages is accompanied by a financial leap that requires small farmers to use savings and/or credit at an unaccustomed scale. Given that small farmers generally have no savings and are already in debt, adopting HYVs entails a relatively higher level of economic risk. Large farmers can commit part of their holdings to new technologies which, if they fail, are backed up by production using previous technologies. If they succeed, there is increased income for debt service and expansion of operations involving the new technologies. For small holders and tenants, commitment to new technologies risks survival or long-term indenture. Large farmers also may have the luxury of withholding their harvest from the market until prices rise again after the post-harvest glut, whereas small farmers usually must sell immediately in order to service debt. Moreover, access to credit and inputs may be dominated by local elites who are also the richer farmers. The rural poor therefore usually have inferior economic and political bargaining positions relative to their richer competitors. (Hardin [1978c] describes the market mechanism as a process in which the power of competitively superior operations increasingly replaces market forces as a determinant of people's ability to compete.)

Pearse argues that as commercialization of the food supply increases, local communities become increasingly dependent on the "urban-industrial network." He claims (198:159) that this form of "incorporation" discriminates against rural people "whose assets include traditional knowledge..., and whose energies are absorbed by the labours of husbandry" in favor of local elites "who have the experience and social attributes necessary for...manipulating the rural-urban nexus." Pearse further claims (1980:181) that the "majority of small cultivators...[believe] that their life situation could not provide the necessary conditions for successful entrepreneurship in view of the known handicaps of poverty and the unknown hazards of the technology itself and of the external dependence to obtain supplies." External dependence in turn reduces small cultivators' decision-making autonomy by linking local production "to national and international episodes of politics and trade and the repercussions of distant war" (Pearse, 1980:160; for similar views of the incorporation of small-scale agriculture into the world economy, see de Janvry and LeVeen [1986], Perelman [1977], and George [1977]. See Frank [1969], Cardoso and Faletto [1979], and Amin [1976] for more general discussions of dependency.)

D. Barkin, R.L. Batt, and B.R. DeWalt claim that the commercialization of Third World agriculture has severe impacts on poor people's access to food. In *The Global Substitution of Grains in Production: Feed Crops for Food Crops* (1990), they examine shifts that occur as grain production becomes increasingly concentrated on larger, more heavily capitalized farms. Barkin et al. note that cash grain crops increasingly displace traditional subsistence food crops as cultivators attempt to take advantage of profitable export markets for grains, forcing poor people into a cash market for food. However, they claim that in the market, poor people must compete not only with affluent foreign and domestic consumers; they also must compete with commercially raised livestock because much of the grain on national and international markets is sold for feedlot and ranch consumption rather than for direct human consumption.

As a result, nutrition standards for the Third World poor become increasingly marginalized because opportunities to earn incomes to enter the cash food market are generally grossly insufficient. (See Lappe and Collins [1979] and Dewey [1981] for discussions of nutritional impacts that result from commercialization of food production.) Moreover, as production for domestic consumption falls, Third World nations become increasingly dependent on imports to meet basic food needs. (See Franke and Chasin [1980] for an account of the impact of the collapse of global peanut markets on post-colonial Sahelian Africa.) They claim that unless people realize the effects of the international market on marginal subsistence farmers, it will be extremely difficult to direct research to the needs of poor people.

A special class of social distribution problem involves the impact of technology packages on the economic and political status of women, especially in the poorest sectors of Third World populations. In "The Underside of Development: Agricultural Development and Women in Zambia," (1985; see Flora 1985), A. Spring and A. Hansen note that women perform much of the agricultural labor in the Third World, especially in Africa. They also note that development plans often focus on men under the assumption that men are the commercial farmers. However, Spring and Hansen argue that although women are often active in commercial agriculture, they may be largely excluded from credit and educational programs that would enable them to take advantage of innovations. Moreover, the introduction of new commercial crops places women at a financial disadvantage in areas where food crops traditionally are grown by women, and cash crops are grown by men. Still further, in a transition from shifting agriculture to a pattern of continuous cultivation, the labor men provide for field clearing is reduced or eliminated, while the labor women provide for cultivation and pest control increases. Spring and Hansen argue that unless planners become more aware of such differential impacts, the already precarious position of rural Third World women will worsen.

Contrasts in the Evaluation of Research

The preceding literature review highlights several issues raised in Ruttan (1987). The advocates of both sides can be construed as responding to the need to tailor research to a specific region's physiographic, socio-cultural, and economic characteristics, and writers on both sides agree that innovative research strategies are needed to address the needs of the Third World poor. Borlaug, for example, argues that the scientific enterprise that produced the Green Revolution must be extended to address farmers in more marginal environments, and that the education process must be broadened to produce integrators rather than disciplinarians. Hayami suggests that research should place more emphasis on adapting HYV packages for contexts of high population growth. Pearse, Barkin et al., and Dahlberg call for a reorientation of research toward subsistence-based strategies.

However, there are differing perspectives on how to analyze the success or failure of research adapted to Third World contexts. Advocates of markets and Green Revolution technologies tend to view success and failure in terms of categories such as total yields, total acreages, and monetary value of national and international trade. They also tend to view social, demographic, and institutional features of Third World societies as factors external to agricultural research and the diffusion of new technologies. One indication of this is Hayami's characterization of the impact of population growth as something apart from the impact of adopting HYVs.[1] The major assumption regarding justice in the *laissez faire* approach is that an ability to participate in free markets is crucial to eliminating rural poverty.

Critics of markets and HYV packages, on the other hand, tend to view success or failure in terms of the differential impact of new technologies on members of different social strata. They view the social distributions of benefits, costs, and economic and political power as factors internal to agricultural research. Their major assumption is that industrializing agriculture leads to dependence on global input and commodity markets, which in turn reinforces domestic inequalities and economic/political domination by First World corporations and governments. The literature therefore suggests that one's view of the success or failure of agricultural development depends largely on the empirical frame of reference used to evaluate it (see Jacobs, 1984; Horowitz, 1972; Hirschman, 1981).

An issue that illustrates the above contrasts is the current dispute over plant germplasm resources (PGRs). Since Pat Roy Mooney's *Seeds of the Earth: A Private or Public Resource* was published in 1979, debate has raged over the ownership and distribution of benefits derived from PGRs obtained from the Third World. Jack R. Kloppenburg Jr., in *First the Seed* (1988a), argues that whereas the industrial North is "gene-poor" and dependent on PGRs from Third World contexts for both its present varieties and future needs, the underdeveloped South is "gene-rich" and recoups little of the net benefits of its contributions to the productivity of seeds marketed by Western corporations. He notes that the developed nations maintain legal and political structures to obtain germplasm freely from developing nations while simultaneously making improved varieties available only on a commercial basis. The industrial nations defend their position by claiming that the costs of using PGRs to improve varieties are high, and that new varieties cannot be generated without commercial opportunities arising from plant breeders' rights over them. (See Brown [1988] for such a defense.)

However, Kloppenburg claims this position unfairly favors commercial seed interests by compensating them for their labor and investment without providing equivalent compensation for Third World farmers who developed and maintained the varieties that serve as the basis for improved proprietary seed lines. (In a similar vein, but from a different perspective, MacKenzie [this volume] notes that some American farmers object to "paying twice" for the results of research.) Moreover, Kloppenburg claims, seed markets are dominated by national and transnational corporations, with the result that farmers in both the First and Third Worlds are vulnerable to exploitation via a loss of genuine choice in the array of varieties (and, in some cases, the other inputs) they can use.

The essays in Kloppenburg's *Seeds and Sovereignty* (1988b) demonstrate the complexity of the PGR controversy. Kloppenburg and D.L. Kleinman suggest in "Seeds of Controversy" (1988) that germplasm should be regarded as a form of national property, and that nations which are contributing PGRs to improved varieties (including those held by First World corporations) should be compensated for the value of their contributions. David Wood, in "Crop Germplasm: Common Heritage or Farmer's Heritage?" (1988), suggests that the Kloppenburg/Kleinman approach could be self-defeating by requiring royalty payments among poor Third World nations. He proposes instead that germplasm should be freely available, but that seed companies should contribute part of their income to a fund that benefits farmers and contributes to maintaining PGRs.

M.S. Swaminathan argues, in "Seeds and Property Rights: A View from the CGIAR System" (1988), that germplasm should be the common heritage of humankind, and control over genetic resources should be vested with the IARCs, which in turn will ensure that Third World nations have access to PGR collections by allowing bona fide researchers free access to collections. In "Molecular Biology and the Protection of Germplasm: A Matter of National Security" (1988), Daniel J. Goldstein claims that because the developed nations dominate the IARCs politically, economically, and in terms of intellectual orientation and scientific technology, Third World nations must take control over germplasm research for their own needs. Jack R. Harlan, in "Seeds and Sovereignty: An Epilogue" (1988), rejects Goldstein's approach and reasons, but emphasizes the overriding importance of building Third World scientific capacity to perform germplasm research and maintain PGR collections. Harlan claims (1988:361) that "Free exchange [of germplasm] provides the greatest good for the greatest number."

The essays on PGRs reflect the conflict between free-market and dependency views of development, as well as differing views on the relevance of unequal distributions of economic and political power to agricultural research and research policy. Considered together, however, the particular issues raised by Swaminathan, Goldstein, and Harlan in turn raise the issue of the value of research entrepreneurship as a basis for maximizing utility. For research entrepreneurship to be a mainspring of research policy for just development, then a central question is how well entrepreneurship can be achieved in agricultural research for developing nations.

Agenda-Setting for Just Development

If, as Ruttan claims, the agenda-setting process is properly a political process, then Dahlberg's comments about seeking solutions that do not reinforce existing inequities pose a profound challenge for research administrators. Ruttan suggests that the research agenda must reflect a balanced set of interests. However, within Ruttan's guidelines, the research administrator's mandate to allocate resources where they can be used most effectively ultimately may be self-defeating if innovations for small farmers require research with long start-up times. As de Janvry and Dethier note (1985), returns on investment in new research streams may accrue much more slowly than returns on continued investment in already well-developed technologies.

However, if de Janvry and Dethier also are correct in asserting that the economic and political powerlessness of poor people keeps them out of the agenda-setting process, then it may be difficult for a research administrator to put their needs on the agenda and keep them there if more powerful interests can use their leverage to allocate scarce resources to mainstream research with a proven track record. Moreover, if research entrepreneurship is vitally important for innovations in agricultural research (as Ruttan claims), then before new research streams can find a place on the agenda, they first must be recognized (or at least tolerated) as legitimate alternatives by members of the established scientific disciplines.

An area in which this last point has emerged is the role of indigenous knowledge in agricultural research. Partly in response to criticisms of Green Revolution strategies, farming systems research (or, FSR; see Norman, 1979; Byerlee et al., 1980; Shaner et al., 1982) and agroecology (see Altieri, 1987; Gliessman, 1990) emerged as "bottom-up" forms of research that give poor farmers a participatory role in determining the content of the research process. These multidisciplinary approaches to agricultural research identify poor farm families' needs, and develop technology and management packages tailored to their physical, socio-economic, and cultural environments. For example, Bruce Horwith et al., in "The Role of Technology in Enhancing Low-Resource Agriculture in Africa" (1989:70; see also Office of Technology Assessment, 1988), claim that "No viable alternative to low-resource agriculture exists in much of Africa today." They suggest that research should focus on attaining incremental growth based on existing agricultural practices because the varied socio-cultural and physical environments of Africa are unlikely to be conducive to conventional development strategies. (See also Altieri, 1989; Lele, 1984; Eicher, 1984.)

Because of their focus on farm families in their contexts, FSR and agroecological approaches rely on indigenous or local knowledge to identify farmers' perceptions of their own problems. Christina Gladwin, in "Indigenous Knowledge Systems, the Cognitive Revolution, and Agricultural Decision Making" (1989), notes that a knowledge of farm-ers' decision-making procedures, including their perceptions of cost benefits and risks, helps avoid developing technologies that will not be adopted while simultaneously providing a means to identify constraints in the existing production system. Constance McCorkle, in "Toward a Knowledge of Local Knowledge and Its Importance for Agricultural RD&E" (1989), suggests that scarce research resources can be saved by drawing on the knowledge of local people. She further suggests that the process of obtaining an understanding of local knowledge brings with it an understanding of local information distribution systems that can serve as the foundation for extension, a crucial element in the bottom-up approach.

In "Power, Class, and Traditional Knowledge in Sahel Food Produc-tion," Richard Franke (1988) documents several cases of USAID studies that draw on integrated field studies of traditional knowledge and practices, but ignore those aspects to focus on disciplinary technical questions. He views this as a serious shortcoming which leads to research and development proposals that aggravate rather than solve food production problems. He claims (see also Altieri, 1983) that traditional knowledge and science can be highly complementary, and can lead to truly creative and adaptive innovations.

Lori Ann Thrupp takes a different approach. In "Legitimizing Local Knowledge: From Displacement to Empowerment for Third World People" (1989), she notes that incorporating local knowledge systems into research is essential for full participation of farmers in the development plans that affect their lives. However, she claims that a major obstacle to such incorporation is scientists' and development experts' low esteem for the legitimacy of knowledge that has not been derived through laboratory tests. Although in many cases (as Franke claims), local technical knowledge can be validated in the laboratory ("scientized," in Thrupp's terms), she argues that this process should be approached with caution. On one hand, scientizing local knowledge to tell farmers what they already know may strike them as nonsensical. On the other hand, scientizing local knowledge may strip it of its social, religious, and other related value-content. If scientized knowledge is then incorporated into development plans that ignore its value-content, rural people may lose control over the terms by which the success of their own development is evaluated. Thus, Thrupp sees local knowledge systems as a basis for empowering rural people to choose development alternatives suitable to their own desires, interests, and values.

If Dundon (1986; see Thompson, Ellis, and Stout, this volume) has correctly identified challenges to disciplinary professional competence as a constraint on innovation, the indigenous knowledge issue highlights a constraint that may emerge in a research policy that accentuates research entrepreneurship. If the structure of the U.S. scientific enterprise is such that alternative approaches are systematically excluded (a consistent theme in the agroecology and indigenous knowledge literature), then entrepreneurship is realizable only within relatively narrow disciplinary boundaries. If this is the case, then as a practice for maximizing utility, the research entrepreneurship legislated into the Hatch Act and advocated by Ruttan falls far short of the ideal that justifies it. As Mill (1859) might suggest, a fair fight is necessary before alternative approaches can be rejected. However, it may be hard for a researcher in an alternative program to get a fair fight (despite his or her position as best-placed to evaluate that program's potential) if the researcher cannot be recognized as a bona fide researcher by mainstream disciplinary scientists. Thus, the issue of who counts as a legitimate researcher directly challenges arguments such as Swaminathan's position on PGRs.

This issue is even more relevant because if Dundon is correct, challenges to professional competence are accompanied by challenges to other values, especially ethical commitments. One of the main criticisms of conventional approaches to research is that researchers have been oblivious to predictable justice-related aspects of the products of their research. Regardless of whether or not this charge is true, scientists face some genuine conceptual challenges in relation to formulating research problems that are compatible with distributive justice. Not the least of these challenges is to identify the conception of justice with which research must be compatible. (See MacIntyre [1988] for a discussion of evaluating a given action from different ethical frameworks.)

Many American scientists, out of duty, charity, or other motivations, direct their research efforts to Third World development projects. Some utilitarians have argued that utilitarianism requires the First World to aid Third World development (in contrast with Hardin's utilitarian judgment that prohibits food aid, if not also other kinds of aid). Peter Singer, in "Famine, Affluence and Morality" (1972), argues that people in affluent nations have a duty to vastly increase the amount of aid they give because they can do so without hurting themselves in any morally significant way. Singer admits that working out the formula for aid is a difficult affair, but believes that the first step in any event is to abandon the notion that aid is properly a form of charity. (See Nagel [1977] for a nonutilitarian argument that reaches a similar conclusion.)

However, within a utilitarian orientation, identification of appropriate innovations is impeded by the difficulty of identifying a standard of the utility to be maximized. Arguments about what counts as "utility" are nearly as old as utilitarianism itself. Mill (1861) argued against Bentham's earlier hedonistic definition of utility. Several different varieties of utility have been proposed since Mill's time (see, e.g., Sen, 1985). Recently, Jeffrey Burkhardt has argued in "The Value Measure in Agricultural Research" (1986a) that defining utility in terms of basic needs is especially appropriate for agricultural research. Thus, even within the presumed utilitarian framework of U.S. agricultural research, a research plan that attempts to accommodate justice must identify the kind of utility to be maximized before it can begin to maximize it.

Onora O'Neill claims in *Faces of Hunger* (1986) that the utilitarian approach to evaluating obligations to the Third World poor is inappropriate. She argues that utilitarianism's directive to maximize utility is insensitive to the actual distribution of utility.[2] She also suggests that utilitarianism flounders because the ability to make ethical judgments depends on the empirical framework the utilitarian uses as a foundation for normative analyses. Utilitarian ethical choices depend on the ability to predict the outcomes of alternative acts or policies. O'Neill claims that because one's theoretical perspective determines what counts as a fact (and, hence, as a successful or failed prediction), and because theoretical perspectives are essentially value-laden (see also Madden, 1986), any ethical judgment based on the factual nature of a set of consequences is partially determined in advance. O'Neill suggests that the defined mission and orientation of development institutions further shape the terms under which the ethical and practical value of alternative development programs will be assessed. (For example, Shaner et al.

[1982] note that achievement of procedural goals—e.g., number of tests performed—can be confused with success of a program's development goals.) Thus, a research policy that targets a specific kind of utility has already eliminated some kinds of facts from consideration in an ethical evaluation of the impact of research.

Some authors suggest that just development is hindered by commitments to particular analytical concepts of the role and nature of nation-states. Charles R. Beitz's *Political Theory and International Relations* (1978; see also Jacobs, 1984) argues that international justice has been inhibited historically by the erroneous assumption that nation-states are autonomous, self-sufficient entities with boundaries that define the limits beyond which there are no obligations to achieve distributive justice. (This claim, if correct, challenges a basic premise of "lifeboat ethics.") Beitz suggests (1978:123) "that the rapid development of [Third World] societies requires a new international economic order as well as changes in [internal] social and political institutions....Analogously, the development of just domestic institutions in many societies may depend on the elimination of international distributive injustice."

Beitz argues that Rawls' (1971) theory of the just social contract can be extended to relations between nations to define the nature of international distributive justice. Oversimplified, Rawls' theory claims that people would not contract into a society that would systematically allow their position to be made worse off while other groups' positions are systematically favored. Thus, for Rawls, any distribution of the benefits of society that raises the share of better-off groups while lowering the share of worst-off groups is an unjust distribution. Central to Rawls' position is the notion that the social position one is born into (among other things) is unearned and arbitrary, and, hence, morally irrelevant to distributive justice. Similarly, Beitz claims that political boundaries are historically arbitrary, and, hence, morally irrelevant to international distributive justice with respect to benefits derived from natural resources. As a result, Beitz claims, members of poor societies have a morally justified claim to portions of rich nations' distributive shares to the extent that those shares are derived from arbitrarily distributed natural resources.

Beitz's claim, although based on a nonutilitarian social contract, is roughly analogous to utilitarian judgments (such as Singer's) which require redistribution of wealth to the Third World poor when such redistributions would fulfill the utilitarian maxim. However, Beitz's position diverges from utilitarianism because utilitarianism might require no redistribution, redistribution to wealthy Third World groups, or even redistribution to the First World if any of these options would lead to an optimal outcome. Furthermore, Beitz's position, if justified, has interesting implications for the land-grant system: Because the land-grant institutions, by definition, derive part of their endowments from their respective states' natural resource bases, Third World peoples may have a morally justified claim to some of their research. If Third World peoples do have such a claim, also at issue is whether they have a claim to influence the research agenda at those institutions.

Billie DeWalt's "The Cultural Ecology of Development: Ten Precepts for Survival" (1988b) suggests that both capitalists and socialists have focused development on strategies aimed toward the industrialization of Third World nations. He claims that the industrialization approach is ecologically destructive in terms of long-term human adaptation, and emphasizes the means of increasing production without a corresponding emphasis on the kind of society development is supposed to lead to. DeWalt's approach seeks to develop sustainable human systems as frameworks for the full development of human cultural, intellectual, and adaptive capacities rather than to develop economies as frameworks for increasing levels of wealth and consumption. (Along this line, Goulet [1973] and Crocker [1990] see Western industrial societies as underdeveloped in the sense that they foster mass consumerism rather than authentic personal development.) Moreover, he claims, it is necessary to focus research and development efforts on ecoregions rather than nations in order to accommodate development to natural and cultural systems that operate independently of political boundaries. Thus, although it is nonutilitarian, DeWalt's position is analogous to Mill's (1859) utilitarian position in that DeWalt sees cultural diversity and self-directed personal development as virtues to pursue in the interest of long-term human adaptation.

DeWalt's position is closely related to several autonomy-based approaches to international justice. These approaches, such as Denis Goulet's *The Cruel Choice* (1973; see also O'Neill, 1986; Crocker, 1990), suggest that development plans which offer poor people a choice between no assistance and integration into one or another global structure do not offer much of a choice because to opt out of integration is to opt into starvation or a degraded lifestyle. This kind of choice is said to violate the autonomy of the poor in the sense that they are essentially not free to reject it, and accepting it entails working for First World interests, whether those interests are compatible with their own or not.

In this approach, then, even a development plan that raises overall utility by incorporating people into a global system in which they would rather not participate is unjust because it treats the poor as means to an end they might not choose if their circumstances were not desperate. FSR and agroecology rely (perhaps not exclusively) on a nonutilitarian, autonomy-based conception of justice which recognizes (as did Mill) that the individuals who will ultimately benefit or suffer must be able to choose which portions of traditional and scientific programs are appropriate for pursuing their goals. Thus, the issue in autonomy-based approaches is the availability of genuine choice. Similarly, positions such as DeWalt's do not preclude choosing First World style development if Third World populations identify it as the appropriate means of achieving their goals. This implies that there is no necessary connection between autonomous and sustainable development. However, if development toward global industrialization is maladaptive, as DeWalt claims, it is still an open question whether such development is moral or immoral on some grounds other than justice (see Kennan, 1985).

Concluding Remarks

The preceding discussion suggests that distributive justice challenges the utilitarian/research entrepreneurial foundations of U.S. research policy in at least two related arenas. The first is philosophical, and involves the compatibility of utilitarian values (as expressed in productivity-oriented research and technologies) with the value of liberty of choice. To the extent that agricultural research develops technologies with value that requires entry into global input and commodity markets to the exclusion of technologies that do not require it, then research provides poor farmers with a choice between enduring or deepening poverty, and a commitment to participation in the global economy. This dichotomy provides no options for farmers who might prefer a choice between starvation and self-provisioning, or for nations which might prefer not to tie their development too closely to fluctuating global markets. The issue for many nonutilitarians is not whether opting out of the global economy is a good idea. Nor is it whether poor Third World people desire First World lifestyles. The issue is whether the available choices genuinely allow the Third World poor to determine the development courses they will take as individuals and nations.

If research policy is such that it does not make genuine choice available, then research policy-makers have effectively decided for the Third World poor what their range of options will be. This not only effectively negates the political goals of persons who do not happen to value membership in growth- and consumption-oriented economies; it is also antithetical to the notion of liberty of choice, whether such liberty is construed as a paramount utility in utilitarianism, a nonutilitarian constraint to utilitarian judgments, or a foundation of nonutilitarian concepts of justice. Indeed, some of the more blistering attacks on conventional development (e.g., Perelman [1977] and George [1977]) begin with extended discussions of U.S. agricultural development, and ask why we would want to inflict a similar future on the Third World. Along parallel lines, Kennan (1985:216) implies that until the U.S. resolves its environmental and fiscal dilemmas, it is not in a position to suggest (however overtly or covertly) how other nations should run their affairs: "in world affairs, as in personal life, example exerts greater power than precept."

The second arena is scientific, and involves giving alternative research programs a fair hearing in the agenda-setting process. Scientific freedom within the established disciplines is based on the assumption that open debate on a wide spectrum of ideas produces the knowledge needed for adaptation to different circumstances. Because the experience of people (including scientists) varies, only those who are closest to a given set of circumstances (including a given research stream) are well-placed to evaluate a proposed course of action. The value of research entrepreneurship resides in providing a context in which people can accept or reject knowledge claims as they see fit given the course of debate and practice. Genuine research entrepreneurship therefore requires expanding the agenda to include programs that may not be recognized as scientifically legitimate by mainstream disciplinary researchers. If the high value placed on research entrepreneurship is justified, practice and debate will demonstrate which of the mainstream and alternative programs will survive, and which will not.

It is in this sense, then, that the two arenas are related. Liberty of choice cannot occur unless there are alternatives to choose from. Practice and debate cannot demonstrate the relative utility of alternative programs if the alternatives cannot get on the agenda. And alternatives cannot get on the agenda if the people who set the agenda are not well-placed to judge the alternatives' potential because the alternatives fall outside their range of experience. Advocates of the value of indigenous knowledge therefore pose a fundamental challenge to the role that disciplinary scientific thought plays in the empirical evaluation of proposed research. Similarly, the nonutilitarian concepts of justice pose a fundamental challenge to the notion that a focus on productivity-oriented research is consistent with the liberty of choice on which research entrepreneurship and other democratic values are based.

If this essay has been moderately successful, it demonstrates that the justice-related positions expressed in the literature can be differentiated in terms of the variables they regard as empirically relevant to agricultural research, and the variables they regard as ethically relevant to the impact of research. Within any given position, these two sets of variables are closely related and mutually reinforcing. The literature therefore suggests that at least part of the conflict in discussions of agricultural development and distributive justice is based on mutual misunderstanding, which in turn may be based on commitments to divergent empirical and ethical frameworks. This essay concludes with an observation that should be considered by all who have input into the agenda-setting process: It is easy for members of different camps to talk past each other because it is difficult to appreciate another's position when it is framed in terms of assumptions and values that differ from one's own.

Notes

1. In a more subtle indication of this, Schuh and Tollini (1979) discuss income distribution, employment, and environmental consequences in a chapter called "The Side Effects of Technical Change." Hardin's (1987c) definition of "side effect" as something we did not foresee or do not want to think about is especially interesting in this context.

2. See Nozick (1974) for a *laissez faire* criticism of this aspect of utilitarianism. His argument further claims that because an unbalanced distribution can result from free, just exchange of goods, redistribution under utilitarian principles negates individuals' rights to dispose of justly acquired property as they see fit. Nagel (1977) argues that even if current global distributions were not largely based on unjust transfers of wealth during conquest and decolonization, the First World would still have a duty to use its wealth to alleviate Third World poverty.

12

Beyond the Large Farm

Bill A. Stout and Paul B. Thompson

Introduction

What is the value of the family farm? Is there a legitimate role for "family farm research" today? What are our duties to future generations, and, indeed, to the natural world itself? Are the reductionist research methods of agricultural science responsible for environmental degradation? What is our duty to the foreign poor? Is American agricultural science universally applicable to agricultural production systems across the globe?

These are among the questions the *Beyond the Large Farm* project has attempted to address. They have been discussed in this volume in the series of chapters written by our contributors, each of whom brings a personal history and a unique point of view to the evaluation of research goals for agriculture. Within the activity of the Texas A&M University "core group" of scientists who participated in seminars, discussions, and workshops during a 2-year period, we attempted to address key questions from the perspective of agricultural scientists working at an institution devoted to agricultural research. While each member of the core group came away from the project with perhaps a renewed personal perspective, we knew at the outset that a small group such as ours could not have a significant effect upon an institution as large as Texas A&M University, much less the USDA/land-grant system. While each member of the core group grew to better appreciate the ethics of agricultural research, we concluded the project with a profound sense of its limitations. There are too many scientists with too little time to devote to the intense reflection and discussion that was experienced in the core group for efforts like this one to have a measurable effect.

While we never expected to achieve total consensus on the values that should guide agriculture, we sought to understand (and to be able to communicate) how values influence what agricultural scientists do, and how engagement in dialogue about the value of agricultural research might influence future practice. In the introduction to this publication, we (with Lain Ellis) used the concept of the research laboratory to analyze the problem. Although the laboratory is equipped with machines and supplies, in placing the laboratory under the direction of scientists trained in specific disciplinary traditions, it is equipped with values, as well. We described these values as functioning within the laboratory to establish criteria for evaluating the scientific quality of laboratory activity (e.g., Does the laboratory produce good science?), and to establish a vision of which research is most worthwhile in relation to broader societal goals.

Although the laboratories of agricultural research vary tremendously, the laboratory analysis of research policy suggests a generic version of the bench scientist's perspective on research choices and products. Biochemists and engineers bring significantly different values to their labs, and differences are reflected in the varied ways in which each lab is equipped. Such differences notwithstanding, USDA/land-grant scientists have all tended to include similar beliefs about productivity and increasing yields among the equipment in their labs. Productivity, then, is a value operating generally within agricultural laboratories, not merely an idiosyncratic value of specific scientists.

Each of the authors who contributed to this book has offered alternatives to the value of productivity. Vernon Ruttan offers what might be regarded as the most modest account of alternatives. He sees emphasis upon increasing biological yields continuing to serve as the central focus of agricultural science. The alternative values noted by Ruttan—health and employment, to name two—must be factored into research planning activities. But society will prosper if scientists pursue their activity in an environment relatively free of external influence.

Miguel Altieri offers the most radical alternative to productivity, for he believes increases in yields play a minor role when compared to the social and environmental goals of agricultural production. Furthermore, he finds that the scientific values of the laboratory—the emphasis upon replicability and scientific controls—spill over into agricultural technologies in ways that often produce socially harmful consequences. Ruttan believes that the main problems arising from agricultural science originate outside the laboratory, with research administrators, funding agencies, and public officials who neglect responsibilities, while being overly responsive to political influence. Altieri thinks that serious problems originate within the laboratory, with the values scientists apply in making judgments about objectivity and explanatory relevance. However, he also maintains that agricultural science cannot solve the current problems in our food system.

Most of the other authors in this volume take the position that scientists should modify the values at work within their laboratories to some extent. Like Ruttan, David MacKenzie wants scientists to be able to pursue their work in relative freedom, but he notes that the agrarian values that a generation of scientists brought into their laboratories from their own family farm life can no longer be assumed. Although MacKenzie is reluctant to suggest specific alternatives, he endorses the need for scientists to address ethical issues in their research activity. Jeffrey Burkhardt also stresses the need for scientists to be cognizant of ethical values when making research decisions, placing particular emphasis upon justice. Like Ruttan, Burkhardt believes that matters of social ethics will be resolved at the level of public policy, rather than within the scientist's research activity.

Norton, Esmay, Cardwell, and DeWalt identify specific changes that must take place in laboratory practice for ethical issues to be addressed. Bryan Norton describes a conceptual shift, a change in our way of thinking that takes place when we begin to perform scientific research from an ecological frame of mind. Norton believes that this shift is less an abandonment of utilitarian, outcome optimizing ethics than it is a transformation of scientific values, or what counts as a complete and adequate scientific explanation. If Norton is correct, development of hierarchy theory in biology will advance the cause of incorporating ethics into agricultural research. Vern Cardwell notes a similar shift of consciousness, but rather than emphasizing conceptual models, Cardwell believes the move to systems methodology will help effect the change. Like Norton, Cardwell thinks that the problem is less with the underlying utilitarian ethic of agricultural research and more with research techniques that insist upon understanding each component of a food system in isolation from the whole.

Merle Esmay notes a slightly different version of the problem, one which becomes obvious in developing world agriculture. Western research methods and technologies are often inappropriate in developing countries. False standards of scientific quality reinforce a human tendency to impose a single solution for every problem; but insistence upon Western criteria of objectivity and proof defeat attempts to solve problems in Asia, Africa, and Latin America. Esmay seems to be saying that the scientist should be a problem-solver, and that single-minded adherence to the Western rules for scientific research and education is foolish when it prevents finding solutions to problems. DeWalt makes an even stronger statement of the same point, claiming that agricultural scientists' insistence in working on conventional yield enhancing research programs has produced a series of predictable (and probably avoidable) disasters in the developing world. Like Cardwell, DeWalt stresses a systems research methodology, one which requires close coordination among scientists from a variety of scientific disciplines. Problems will recur until scientists find a way to conduct genuinely collaborative research; explicit attention to ethics may be an avenue for interdisciplinary cooperation in planning and monitoring research activities.

For the most part, these authors have concentrated on internal values, values that operate within the laboratory and are understood to represent a part of the laboratory's practice *qua* laboratory (see Rudner, 1953). The bibliographic essays in this book survey a different type of value influence, however. They illustrate that the critics of agriculture, from Rachael Carson to Wendell Berry, have become a force (although a weak one, in many cases) that attempts to impose values upon the laboratory. Through political action and influence upon administrators and funding sources, critics have attempted to affect what can and cannot be accomplished in agriculture. Although critics usually target agriculture, rather than agricultural research specifically, the values that underlie their claims constitute an external force imposed upon agricultural scientists. As noted, it is a force that has often proved resistible, particularly in the arena of international development issues, but it is a force, nonetheless. Responsible agricultural scientists will want to make rational and relevant replies to critics' points.

Because the values that support criticism of agriculture come from outside the mainstreams of thought within agricultural sciences, we call them external values. The criticisms of agriculture that are chronicled in the initial pages of this book arise from external values. Although the revision of internal values, which has been the primary focus of discussion during the first eight chapters, clearly has been prompted by the external critics, it is important to conclude the volume with a return to the give and take of external values. The book summarizes with some sense of where agriculture stands with respect to broad societal trends. External criticism has many facets; criticisms are offered from a variety of viewpoints on the issues. One vantage point is certainly that of the critics themselves, a perspective that was reviewed at length in the three bibliographic essays. Equally important is the perspective on criticism when examined from inside the laboratory, or the view of the practicing agricultural scientist. Thus far, the book has focused on the scientists' view of their own laboratory activities, but now it is time to devote some thought to their view of the critics themselves. We call this perspective on external values *The View from the Bench*.

A third perspective, different from that of the critics and the scientists themselves, is that of society—or perhaps, history. How does this controversy look to a complete outsider, to someone who views critics and scientists as competing interests? How does the controversy appear from a position far removed from the practical issues of political action (the practical activities that consume the critic) or scientific research (the practical activity of the laboratory)? Of course, there is no one in such a position; no living human being's interest in agriculture is entirely abstract. So, because no one truly has such a view on agriculture, we have chosen to call this perspective *The View from Nowhere*, after the title of Thomas Nagel's 1986 book. The view from nowhere is an abstraction that tries to approximate the perspective that future historians of agriculture will have. Many have claimed that agriculture truly is contributing to the great progress of our society (or to its demise). In our own time, such claims serve as praise or condemnation, while for a future historian, they are hypotheses in an attempt to understand the transect of human society. The view from nowhere is an attempt to move beyond praise and condemnation, and to approximate a vantage point that we, admittedly, cannot truly attain.

The View from the Bench

The purpose of agriculture is to produce food. Producing food is an energy conversion process. Producing food provides us with employment and income, and thus a wholesome place to live. Primitive agriculture consisted of scattering seeds on the land and accepting a meager yield. Little or no "cultural" energy was involved. Cultural energy is all the direct and indirect energy used to produce a crop, other than solar energy. It includes fuels, fertilizers, and pesticides. The result of primitive agriculture was a high energy ratio, but a low yield. Solar energy conversion to food energy was extremely low.

Gradually, humans learned that organic matter and simple fertilizers would increase yields. They also learned to use animal power to expand the area cultivated and reduce human effort. The consequence of this transition was increased yields, but decreased energy ratios. Through the establishment of the land-grant university system and other research institutions, modern scientific agriculture evolved, with less labor input and more cultural energy input. Extensive use of fertilizers and pesticides has led to the significantly increased yields that are necessary to feed a growing world population. The increased cultural energy input, however, has resulted in a lower energy ratio.

Any discussion of alternative production systems must recognize constraints upon the conversion of energy to produce food. Whatever else they are, human beings are a potential source of energy for this process, but their potential contribution is quite limited. Many studies have documented the maximum human power output on a continuous basis at about 1/10 hp or 75 watts. If a human continues this power output for a 10-hour workday, the result is 750Wh or 3/4 kWh. At today's electrical rates, the value of this amount of energy is but a few pennies per day. Humans simply are not effective sources of mechanical work, and can never be competitive with other power sources. Examined another way, consider the work output of a human compared to food energy intake. Suppose a person eats 3,000 kcal of food energy each day. This amount of energy converts to 3.5 kWh, which is four or five times the work energy capacity of a human being. Therefore, humans must supplement their work capacity with other forms of energy to survive.

Modern agriculture is an evolved system for supplementing human energy available for the production of food. In the past century, the primary avenue for advancement in this system has been through narrowly defined, reductionist research aimed at specialized areas. Agricultural researchers are good at this type of research. After more than 100 years of agricultural research, yields are high, labor input is low, input costs are high, chemical use is high, and environmental problems persist. Food is abundant, commodity surpluses are common, and gluts exist in many commodity areas. The U.S. has the lowest food-to-income ratio in the world—less than 15 percent of disposable income. Millions of people have left the family farm for what they consider a better way of life. Yet, we have an uneasy feeling that the loss of the family farm constitutes a larger social crisis. Farms are merging, their average size is increasing, and farmers, especially small farmers, are going broke.

These events create real problems, and critics of agriculture are correct to note them, yet it is not clear that these are specifically agricultural problems. Is the loss of family farms a greater crisis than a factory closing that displaces hundreds, or thousands of people? Is it more of a crisis than a small non-farm business that is forced into bankruptcy, or a buggy manufacturer who loses the market as new technology evolves? Romantic myths about family farms only make it more difficult to face other problems that are associated with modern agriculture. Humans have limited energy capacity, and we eat more food energy than we can generate.

U.S. agriculture is energy intensive. Energy-related inputs such as fertilizers and pesticides, irrigation, and mechanization produce high yields with low labor requirements. But high levels of energy use also may lead to serious environmental concerns. Unacceptably high levels of soil erosion may result from large-scale mechanized farming practices if insufficient attention is given to soil and water conservation. Ground and surface water may be contaminated with fertilizers and pesticides. Food supplies may be at risk from residues or chemicals designed to enhance the production of crops or livestock.

One of the messages throughout this book is that increased agricultural production should not be the only, or even the primary goal of the agricultural research system, especially in an era when the supply of many commodities exceeds the demand. Practices aimed at increasing productivity should be weighed against the environmental costs, and a more socially acceptable balance sought.

We live in a competitive world. For example, rice can be grown in Thailand and shipped to Houston at lower prices than Texas rice growers can produce it. Similarly, Brazil can produce soybeans and Argentina can grow wheat cheaper than American farmers. Agriculture is a global industry, and we cannot ignore economic realities and trade balances. The competitiveness of world agriculture means that we must be sensitive to trade offs between profitability and environmental quality.

Sustainability is one of the major issues in agriculture today. But is the family farm more sustainable than other types of farms? We, as a nation, cannot afford to adopt the Amish model of family farming. A multitude of people left the family farm for a better way of life. At that time, many family farms had no electricity. That means no refrigeration, a hand pump in the kitchen, corn cobs for heat in the kitchen stove, an outdoor toilet, a weekly bath in a metal tub, oil and gas lamps, etc. No one can sell that lifestyle to the American people voluntarily. It is a romantic myth to many today, but virtually everyone who lived it wanted to escape.

Agricultural research is designed to produce changes in traditional farming techniques. As Dr. Norman Borlaug said at one of our workshops for the *Beyond the Large Farm* project, we are dealing with the psychology of change, not just agricultural science and technology. A new technology (e.g., variety or production system) must increase yields significantly—by 30 percent or more. Change involves risk and the Third World cannot risk unless the potential benefit is significant.

When developing a strategy for introducing improved technologies, Borlaug suggests that a scientist must keep the plan simple. Elaborate plans are not effective in selling government decision-makers. Make sure all the parts are there, he says, and go one step at a time. The most convincing argument for a new technology is a successful large-scale, on-farm demonstration.

One cannot assume that good research will produce more food, especially if the research addresses only one dimension of the problem. An integrated package of research is usually needed, i.e., varieties, disease resistance, improved agronomic practices, methods of planting and seed spacing, proper placement of fertilizers, restoration of soil fertility, conservation of moisture, weed control, irrigation water application rates and timing, and so on. Research must be accompanied by policies that provide incentives to local growers. Commodity prices must be attractive, allowing growers to make a profit.

For world food needs to be met, chemical fertilizers and pesticides are an essential imput. Worldwide environmental concerns have led critics to advocate restricted use, or even elimination of chemicals in agriculture. With today's scientific instruments, we have the capability to detect one part per billion—or a fraction of a part per trillion. There is no zero. Low concentrations of pesticides can be beneficial in crop production and not necessarily harmful to humans. But with the ability to detect minute quantities, there is external pressure to avoid all detectable levels of pesticide residues, even levels that may occur naturally. Extreme limits on pesticide use may be a serious deterrent to crop production. Agricultural researchers have lost the general public. The general public, and even college students, know little about agriculture and the importance of appropriate technologies, including chemicals.

Today's agricultural industry represents a serious social and economic dilemma. Research and social scientists, like the authors of this publication, conduct workshops and write books and articles on the problem, but we have not found easy answers. To begin to approach viable solutions, we must look beyond the "productionism" of the large farm. Social, moral, and philosophical issues must take a front seat in the realm of agricultural research. The land-grant system must change, and it is changing. But is it changing in the right direction?

Base budgets for agricultural research have eroded drastically in the last decade. Research in the agricultural experiment stations is now largely funded by grants and contracts. This means that the individual researcher has a weaker and weaker voice in setting the research agenda. Basic research is now the name of the game at major land-grant universities. Some administrators have tunnel vision focused on one aspect of agricultural research, such as biotechnology. Will biotechnology help save the family farm? If so, when? A decade from now? In two or three decades?

The View from Nowhere

In a recent book entitled, *The Condition of Postmodernity* (1989), geographer David Harvey writes about the widespread cultural abandonment of "the project of modernity." The project of modernity is the shared vision of humanity's future (and of the means that industrialized societies would use to reach it) that was dominant among educated people in both capitalist and centrally planned economies from sometime during the 19th century until sometime in the recent past. Paraphrasing the German philosopher Habermas, Harvey describes the project of modernity in the following way:

> The idea was to use the accumulation of knowledge generated by many individuals working freely and creatively for the pursuit of human emancipation and the enrichment of daily life. The scientific domination of nature promised freedom from scarcity, want and the arbitrariness of natural calamity. The development of rational forms of social organization and rational modes of thought promised liberation from the irrationalities of myth, religion, superstition, release from the arbitrary use of power as well as from the dark side of our own human natures. Only through such a project could the universal, eternal, and the immutable qualities of all of humanity be revealed. (p. 12)

Harvey thinks that this vision of the future was largely abandoned by thinking people sometime around 1972. Harvey's book goes on at some length to document critics of the modern view in such diverse areas as architecture, philosophy, business, communications, technology assessment, and literary criticism. The fact that we are preparing this volume nearly 2 decades later is proof that agriculture continues to be slow in getting the news.

Harvey's description of the project of modernity is relevant to this chapter because it also describes the self-image of the agricultural scientist. The internal values of agricultural research assume that science is a means for realizing the betterment of human life. From this perspective, a modern scientist does not really need to ponder the question of what counts as "bettering" human life, because the modern view assumes that science's ability to control events within a food system will serve whichever definition of "betterment" is finally selected. The cultural shift that occurs sometime in the 1970s, however, is marked by a rejection of modernism as a way of thinking. We follow Harvey's terminology and identify the rejection of modernism with the term *postmodernism*. The postmodern critics of society do more than find fault with the products of scientific research. They reject the vision of humanity's future that underlies agricultural research, and conclude that commitment to the project of modernity is the primary source of the problem.

Perhaps a cultural sea change has taken place since 1972, and perhaps not. The discussion of postmodernism helps us arrive at the view from nowhere, in any case, for it shows us that the kind of criticisms leveled against agriculture are not unique to agriculture. In fact, they fit a pattern of criticism that has been repeated in virtually every area of culture. Examined from within the laboratory (from within the project of modernity), environmental and social externalities of agricultural technology simply demonstrate the need to do a better job of applying science to improve humankind. But social, environmental, and international criticisms fit a pattern that reverberates throughout our culture. Without claiming to circumscribe the postmodern world view, we note four themes of conformity to that pattern.

1. **Questioning the possibility of scientific mastery:** Environmental problems in agricultural production systems are often cited as evidence that science is inherently limited to what can be known in a given time frame, while the full complexity of ecological systems can never be known all at once.
2. **Acknowledging the authority of other voices:** Internationalist criticisms of agriculture have repeatedly claimed that scientists cannot assume that elimination of poverty and poor yields are desired by peasants of other cultures, particularly when they also mean elimination of the ways of life that have defined their sense of home.
3. **Advocating participation, rather than optimization of outcomes:** Family farm critics, international development critics, and animal welfare critics all claim that we cannot fail to consider affected interests, even when allowing affected parties to participate in policy and planning ensures that we will sacrifice what might be achieved by beneficial outcomes.
4. **Abandonment of the commodity form:** Systems research in agriculture forsakes the standards that traditionally have been applied in the peer review process in pursuit of research results that are idiosyncratically tailored to client needs. Organic certification, designer foods, custom tillage, and integrated pest management are all examples of technologies that succeed because of their ability to meet highly specific types of client demand that cannot be universalized.

At the same time, if the *products* of agriculture become "decommodified," agricultural research itself becomes a commodity. Abandoning the project of modernity robs agricultural research of its utilitarian justification for science to serve the common good. Mill's account of research entrepreneurship, free, unfettered inquiry leading to truth and the betterment of life, is replaced with an account of market forces and political budget decisions that lead where the rich and powerful want them to go. Time will tell whether scientists who doubt mastery, consider relativism, invite participation, and market their skills to produce designer foods on demand can survive the process of tenure and promotion.

These four themes are common in many areas of postmodern American culture. In architecture, for example, mastery has been questioned because art and architecture critics are no longer considered to possess indisputed expertise in judging the validity and quality of design. The comprehensive theories of design that were once taught in architectural schools are now regarded as relativized points of view or alternative conceptions of design, rather than as principles for mastering design. Abandonment of mastery and flirtation with relativism, thus, go hand-in-hand. Architects advocate participation to the extent that the postmodern school aims to please the people who will inhabit architectural spaces, rather than adhering to formal principles. Clients are often quite literally invited to participate in the process of design to an extent that was unheard of several decades ago. Abandoning of the architectural commodity form—the steel, glass, and reinforced concrete rectangle—is obvious to anyone who has seen the spectacular new buildings going up in every major American city. New buildings compete for occupants, and the market for architectural space has learned to demand unique and peculiar spaces, decorative icons, and visually singular facades (Harvey, 1989).

From the viewpoint of a cultural observer who occupies no particular place among the producers and consumers of agricultural and architectural goods, the coincidence here is striking. It becomes even more striking when the same analysis is applied to medicine, Wall Street, literature, music, theater, and the visual arts. Although it is easy for those of us in the thick of it to be oblivious to such patterns at work in our culture, it is likely that we cannot fully cope with these sweeping winds of change from an isolated point of view. Our viewpoint is preoccupied with specialized projects and defensive of the values that underlie personal and professional interests. *The View from the Bench* is, of course, isolated and guided by internal values in just this way. *The View from Nowhere* is not a perspective that any of us can truly occupy, but to the extent that we can see postmodern society from that vantage point, we can place internal values into a genuine dialogue with external criticisms. We can transcend the constant recrimination and defensiveness of recent years, and, with luck, begin to discuss the agriculture of the next century.

Summary

The agricultural research system was developed to solve specific problems. It is very good at that, but we need to look at the bigger picture. Unquestionably, many researchers should be studying the specifics in depth, but others should be looking at more general issues.

What's wrong with the agricultural research community? It's too damn scientific. It's not broad enough to consider social and ethical issues. Scientists are too far removed from the end use; they are ignorant of their own social context. Specialized language and concepts inhibit interdisciplinary work, to say nothing of dialogue with non-specialists. Specialization is the only way we know how to do science, but agricultural research should include humanistic specialists as well. The land-grant university needs new priorities which include goals such as—sustainability, poverty alleviation, and environmental purity. The entire agricultural system (which includes people) is important in just the way that *The View from the Bench* sees it; but it's hard to deal with social systems all at once. They are too complex, and no one can truly take *The View from Nowhere*.

Have we failed to develop institutions to address general problems in society? Is the land-grant system still too focused on increasing production, even in an era of nagging surpluses? Is the tenure system in land-grant universities part of the problem? Can one achieve tenure as an agricultural researcher by conducting people-oriented systems research? Should we or could we modify tenure rules to allow for less specialization—for some faculty members who want to be professional generalists, integrators who "specialize" in putting the pieces together into a coherent whole for the benefit of people? Could it be that the problem is not at the research level, but at the mission level? Our policy seems to be that saving small farms or solving people problems is less important than maintaining a cheap food supply.

Russell W. Peterson became governor of the State of Delaware and director of the U.S. Office of Technology Assessment after earning a Ph.D. degree in chemistry. His doctoral research topic was "Inhibition of the Air Oxidation of Vitamin C." He said:

> I left (the University of Wisconsin) proud of my new Ph.D. in chemis-
> try—not realizing how little I knew. My lifetime of on-the-job training
> for becoming a generalist has included careers in education, research,
> industrial management, politics, state and federal government, citizen

action, and world conservation and development. I have seen first-hand that it is more the rule than the exception that people enter the most influential positions in our society ill-prepared for the breadth of their assignments. This includes presidents and other policitians, captains of industry, university presidents and deans, cabinet and military officers, judges, editors and journalists, ambassadors, and heads of citizen's organizations. No institution of higher education provides formal training for such important posts.

These jobs (and agricultural policy-makers and some agricultural researchers) call for highly skilled generalists, people who understand such things as the place of Homo sapiens in the biosphere; the interdependence of all people and nations; and how the world's economic, political, social, and management systems operate. Our leaders (and agricultural policy-makers and some researchers) must be able to discern the possible long-term impacts of specific research and development projects. Their decisions should reflect a firm grasp of the historical record from geologic time to the present; the physical, chemical, and biological nature of things; and computerized techniques for assessing trends and appraising alternative futures (Peterson, 1988).

Peterson's remarks stress education. The emphasis in our own project has been research. There is a remarkable tendency to distinguish education and research today, to see the two as separate activities. Yet research is a form of self-education, and conventional classroom education is, for the student, at least a way of "finding out." Research and education build a person's capacity to act, understand, and find out how to act and understand in new and untried circumstances.

The lack of capacity to research and communicate about the food system as a social whole is the predictable result of research and educational choices made decades ago. It is important to recognize that at least two very different forces influenced these decisions: one was scarcity, and the other was the dominant values of modernism, the "project of modernity."

Scarcity of time and money for finding out about everything we might like to know forces us to make choices. The utilitarian values of agricultural research equipped scientists and administrators with powerful tools for setting the priorities that guide such choices. The expenditure of time and money can be evaluated according to a criterion of efficiency: Given a limited amount of time and money, which expenditure will yield the largest social benefit? Viewed in this way, scientists and administrators have examined their own activity just as they have examined the farmer's. Everyone is interested in increasing yields.

Up to a point, nothing is wrong with the utilitarian response to resource scarcity, and getting the most for your effort is little more than common sense. The utilitarian approach to scarcity becomes pernicious when combined with the values associated with modernism, a blind faith that science will solve social problems. By itself, faith in scientists' ability to make discoveries that will benefit humankind is also common sense, a belief that is justified by 2 centuries of progress. When combined with an ethic of responding to scarcity by maximizing yields, however, this belief may seduce us into always choosing the investment of time and money that promises the highest return. If we judge (perhaps correctly) that yields on the expenditure of any particular dollar or any particular day will be maximized when scarce resources are invested in reductionist science, we will never choose to find out about the food system as a whole, much less the social context of science itself.

If we always choose the investment that produces the highest immediate return, we may neglect the shared values and common experiences that make it possible for us to work with one another in the first place. An individual with a successful consulting business may be able to significantly benefit their family by working hard and earning a substantial income. Indeed, it is imperative to the family that income is earned. At any given moment, for any particular 2-hour period, working at the consulting business may make the maximum contribution to family welfare when compared to other options. That is, if the consultant looks at each 2-hour block of time, he or she may ask, "What is the most efficient use of my time?", The answer may always be "consulting." But if this is the decision that is *always* made (even if it is rational from the perspective of efficient use of scarce time resources), common sense tells us that eventually, family welfare will suffer.

At any given moment, we can neglect the food system, the social context of science, and the reflection upon the values that guide our choices. The food system does not require constant monitoring, and during the short-term, the shared values and experiences that allow it to function do operate, whether we pay attention to them or not. If we always choose the immediate efficiencies, however, eventually we produce a generation without shared values and experiences, for the values and experiences that make up the fabric of their lives are wholly constructed from short-run efficiency choices.

We cannot afford to make choices about the allocation of scarce resources by asking only which research project, which equipment purchase, or which use of a scientist's time will result in the greatest increase in knowledge. The knowledge produced by that strategy is without value in a world without community. While we must allocate some of our resources according to an efficiency criterion, we must also allocate some time and effort (and money!) to the construction and maintenance of community, shared experiences, shared concepts, and shared language.

This may seem like a mundane conclusion to some, but it really should come as no surprise that we have failed to offer a technical fix to the ethical problems in agricultural research. No one should expect an effort like our *Beyond the Large Farm* project to produce a technology for making ethical research decisions, and most people would be rightly suspicious if we claimed we had. The concrete result of our project was the little piece of community that was built at Texas A&M University. It was built because 10 extremely busy researchers allocated a portion of their time to a manifestly inefficient use, and because the National Science Foundation allocated precious research dollars to support our workshops, our graduate students, and the production of the chapters in this volume. In itself, the project does not build "enough" community to affect research choices at our university, but community is not easily quantified—even into such rough categories as "too little," "too much," and "enough."

We close, then, by thanking our readers. Those of you who are scientists and administrators have taken some time to read something that probably will not increase your research productivity. We hope that you have been alternately maddened, enlightened, and even bored by what you have read. We hope that you have found some of the ideas in this book wise, some foolish, some radical, and some tradition-bound. We hope that you have smiled when one of our contributors put a thought that has crossed your mind into words, and grimaced when one of them expressed a view that you have entertained and discarded. If you have experienced all of these things, you have experienced the things that every member of our Texas A&M University core group did during the 2-year duration of our project. The points of divergence and disagreement remain, of course, but now dialogue is shared. Although that shared dialogue is only a thread of the community required for wise practice in agriculture and agricultural research, it is a start.

Bibliography

Adams, M.W., Ellingbae, A.H., & Rossineau, E.C. (1971). Biological uniformity and disease epidemics. *BioScience.* 21:1067-1070.

Aiken, W. (1982). The goals of agriculture. *Agriculture, Change and Human Values.* (Haynes, R. & Lanier, R., eds.) Humanities and Agriculture Program, University of Florida, Gainesville, FL. 29-54.

Aiken, W. (1984). Ethical issues in agriculture. *Earthbound.* (Reagan, T., ed.) Random House, New York.

Aiken, W. (1986). On evaluating agricultural research. *New Directions for Agriculture and Agricultural Research: Neglected Dimensions and Emerging Alternatives.* (Dahlberg, K., ed.) Rowman and Allanheld, Totowa, NJ. 31-41.

Aiken, W. & LaFollette, H., eds. (1977). *World Hunger and Moral Obligation.* Prentice-Hall, Englewood Cliffs, NJ.

Allen, T.F.H. & Starr, T.B. (1982). *Hierarchy Theory: Perspectives for Ecological Complexity.* University of Chicago Press, Chicago.

Altieri, M.A. (1983). The question of small farm development: who teaches whom? *Agriculture, Ecosystems and Environment.* 9:401-405.

Altieri, M.A. (1987). *Agroecology: The Scientific Basis of Alternative Agriculture.* Westview Press, Boulder, CO.

Altieri, M.A. (1988). Agroecology: a new research and development paradigm for world agriculture. Paper presented at the International Symposium on Agricultural Ecology and Environment, University of Padua, Italy.

Altieri, M.A. (1989). Rethinking the role of U.S. development assistance in third world agriculture. *Agriculture and Human Values.* 6 (3):85-91.

Altieri, M.A. & Anderson, M.K. (1986). An ecological basis for the development of alternative agricultural systems for small farmers in the third world. *American Journal of Alternative Agriculture.* 1:30-38.

Amin, S. (1976). *Unequal Development: An Essay on the Social Formations of Peripheral Capitalism.* Monthly Review Press, New York.

Anonymous. (1957). Annual Meeting American Society of Agronomy: Resolutions and Necrology. *Agronomy Journal.* 49:603.

Arndt, T.M., Dalyrimple, D.G., & Ruttan, V.W. (1977). Resource Allocation and Productivity in National and International Agricultural Research. University of Minnesota Press, Minneapolis, MN.

Arrow, K. (1951, 2nd ed. 1963). *Social Values and Individual Choice.* Cowles Foundation for Research in Economics at Yale University, New Haven, CT.

ASA/CSSA/SSSA. (1989). *Agronomy News.* Jan.:15.

Asimov, I. (1979). Pure and impure: the interplay of science and technology. *Saturday Review.* 6 (June 9):22-24.

Auxter, T. (1985). Poetry and self-knowledge in rural life. *Agriculture and Human Values.* 2(2):15-27.

Bane, M.J. & Ellwood, D. (1989). One fifth of the nation's children: why are they poor? *Science.* 245:1047-1053.

Banfield, E.C. (1958). *The Moral Basis of a Backward Society.* Free Press, New York.

Barkin, D. & DeWalt, B.R. (1988). Sorghum and the Mexican food crisis. *Latin American Research Review.* 23 (3):30-59.

Barkin, D. & Suarez San Roman, B. (1982). *El fin de la autosuficiencia alimentaria.* Centro de Ecodesarrollo, Editorial Nueva Imagen. Mexico.

Barkin, D., Batt, R.L., & DeWalt, B.R. (1990a). *The Global Substitution of Grains in Production: Feed Crops for Food Crops.* Lynne Rienner Publishers, Boulder, CO.

Barkin, D., Batt, R.L., & DeWalt, B.R. (1990b). *The Substitution of Grains in World Production: Food Versus Food.* Lynne Rienner, Boulder, CO.

Barkley, P.W. & Seckler, D.W. (1972). *Economic Growth and Environmental Decay: The Solution Becomes the Problem.* Harcourt Brace Jovanovich, New York.

Bartlett, P.F. & Brown, P.J. (1985). Agricultural development and the quality of life: an anthropological view. *Agriculture and Human Values.* 2 (2):28-35.

Baum, W.C. (1986). *Partners Against Hunger: The Consultative Group on International Agricultural Research.* The World Bank, Washington, DC.

Bawden, R.J., Ison, R.L., Macadam, R.D., Packham, R.G., & Valentine, I. (1985). A research paradigm for system agriculture. *Agricultural Systems Research for Development Countries: ACIAR Proceedings, No. 11.* (Remenyi, R.V., ed.) 31-42.

Becker, S.L. (1976). Agricultural innovators and innovations, Donald F. Jones and hybrid corn. Lockwood lecture April 9, 1976. *The Connecticut Agricultural Experiment Station Bulletin 763.* New Haven, CT.

Beitz, C.R. (1979). *Political Theory and International Relations.* Princeton University Press, Princeton, NJ.

Bellah, R.N., Madsen, R., Sullivan, W.M., Swidler, A., & Tipton, S.M. (1985). *Habits of the Heart.* Harper & Row, New York.

Benn, S. & Peters, R. (1959). *Principles of Political Thought.* Allen & Unwin, New York.

Bennett, J. (1976). *The Ecological Transition: Cultural Anthropology and Human Adaptation.* Aldine, Chicago.

Berardi, G.M. & Geisler, C.C., eds. (1984). *The Social Consequences and Challenges of New Agricultural Technologies.* Westview Press, Boulder, CO.

Berry, W. (1970). *Farming: A Handbook.* Harcourt, Brace, Jovanovich, New York.

Berry, W. (1972). *A Continuous Harmony: Essays Cultural and Agricultural.* Harcourt Brace Jovanovich, New York.

Berry, W. (1977). *The Unsettling of America.* Sierra Club Books, San Francisco.

Berry, W. (1981). *The Gift of Good Land: Further Essays, Cultural and Agricultural.* North Point Press, San Francisco.

Berry, W. (1985). *Collected Poems: 1957-1982.* North Point Press, San Francisco.

Berry, W. (1987). A defense of the family farm. *Is There A Moral Obligation to Save the Family Farm?* (Comstock, G., ed.) Iowa State University Press, Ames, IA. 347-360.

Bishop, K. (1987). California U. told to change research to aid small farms. *The New York Times.* November 19, 1987:13A.

Blackstone, W.T. & Heslep, R.D., eds. (1977). *Social Justice and Preferential Treatment.* University of Georgia Press, Athens, GA.

Blaikie, P. (1985). *The Political Economy of Soil Erosion in Developing Countries.* Longman, London.

Blaikie, P. & Brookfield, H. (1987). *Land Degradation and Society.* Metheun and Co., New York.

Blatz, C. (1984). Ethical issues in private and public ranch land management. *Agriculture and Human Values.* 1 (4):3-16.

Bodley, J. (1990). *Victims of Progress.* (3rd ed.) Mayfield, Mountain View, CA.

Bonnen, J.T. (1983). Historical sources of U.S. agricultural productivity: implications for R & D policy and social science research. *American Journal of Agricultural Economics.* 65:958-66.

Bonnen, J.T. (1986). A century of science in agriculture: lessons for science policy. *American Journal of Agricultural Economics.* 68:1065-1080.

Borlaug, N. (1986). Accelerating agricultural research and production in the third world. *Agriculture and Human Values.* 3 (3):5-14.

Borrelli, P. (1988). *Crossroads: Environmental Priorities for the Future.* Island Press, Washington, DC.

Boulding, K.E. (1966). *The Impact of the Social Sciences.* Rutgers, New Brunswick, NJ.

Braun, J. von & Kennedy, E. (1988). Effects of technological change in agriculture on food consumption and nutrition: rice in a west African setting. *World Development.* 16 (9):1083-1098.

Breimyer, H.F. (1977). *Farm Policy: 13 Essays.* Iowa State University Press, Ames, IA.

Brown, G.E., Jr. (1989). Contemporary issues: the critical challenges facing the structure and function of agricultural research. *Journal of Production Agriculture.* 2:98-102.

Brown, L. (1989). *State of the World, 1989.* (1st ed.) Norton, New York.

Brown, P.G. & Shue, H., eds. (1977). *Food Policy: The Responsibility of the United States in the Life and Death Choices.* Free Press, New York.

Brown, W.L. (1988). Plant genetic resources: a view from the seed industry. *Seeds and Sovereignty: The Use and Control of Plant Genetic Resources.* (Kloppenburg, J.R., Jr., ed.) 218-230. Duke University Press, Durham, NC.

Browne, W. (1988). *Private Interests, Public Policy and American Agriculture.* University Press of Kansas, Lawrence, KS.

Buchanon, J.M. (1987). The economizing element in Knight's ethical critique of capitalism. *Ethics.* 98 (1):61-75

Burbach, R. & Flinn, P. (1980). *Agribusiness in the Americas.* Monthly Review Press, New York. Bureau of Census. (1975). Historical statistics of the United States and colonial times to 1970. U.S. Dept. of Commerce, Washington, DC.

Burkhardt, J. (1986a). The value measure in agricultural research. *The Agricultural Scientific Enterprise: A System in Transition.* (Busch, L. & Lacy, W.B., eds.) Westview Press, Boulder, CO. 25-35.

Burkhardt, J. (1986b). Agribusiness Ethics: Specifying the Terms of the Contract. *Journal of Business Ethics.* 5 (4):333-345.

Burkhardt, J. (1988). Crisis, argument, and agriculture. *Journal of Agricultural Ethics.* 1 (2):123-138. Burkhardt, J. (1989). The morality behind sustainability. Journal of Agricultural Ethics. 2:113-128.

Burkhardt, J., Busch, L., Lacy, W.B., & Hansen, M. (1986). Biotechnology and food: a social appraisal. *Food Biotechnology.* (Knorr, D., ed.) Dekker, New York. 575-600.

Busch, L. (1980). Structure and Negotiation in the Agricultural Sciences. *Rural Sociology.* 45:26-28.

Busch, L., ed. (1981). *Science and Agricultural Development.* Allanheld, Osmun, Totowa, NJ. Busch, L. (1989). Irony and Tragedy in Agricultural Research. Presidential Address, Agriculture Food and Human Values Society Annual Meeting, Little Rock, AR, November.

Busch, L. & Lacy, W.B. (1981). Sources of influence on problem choice in the agricultural sciences: the new atlantis revisited research. *Science and Agricultural Development.* (Busch, L., ed.) Allanheld, Osmun, Totowa, NJ. 113-123.

Busch, L. & Lacy, W.B. (1983). *Science, Agriculture, and the Politics of Research.* Westview Press, Boulder, CO. Busch, L. & Lacy, W.B. (1984). Agricultural policy: issues for the '80s and beyond. Agriculture and Human Values. 1 (1):5-9.

Busch, L. & Lacy, W.B., eds. (1986). *The Agricultural Scientific Enterprise: A System in Transition.* Westview Press, Boulder, CO.

Busch, L., Lacy, W.B., Burkhardt, J., and Lacy, L. (1990). *Plants, Power and Profit.* Basil Blackwell, London.

Buttel, F.H. (1980a). Agriculture, environment and social change: some emergent issues. *The Rural Sociology of the Advanced Societies.* (Buttel, F.H. & Newby, H. eds.) Allanheld, Osmun and Co., Montclair, NJ.

Buttel, F.H. (1980b). Agricultural structure and rural ecology toward a political economy of rural development. *Sociologiá Ruralis.* 20:44-62.

Buttel, F.H. (1985). The land-grant system: a sociological perspective on value conflicts and ethical issues. *Agriculture and Human Values.* 2 (2):78-95.

Buttel, F.H. (1986). Agricultural research and farm structural change: bovine growth hormone and beyond. *Agriculture and Human Values.* 3 (4):88-99.

Buttel, F.H. (1987a). Review of *Policy of Agricultural Research* by Ruttan, V.W. & Pray, C.E. Westview Press. *Agriculture and Human Values.* 4 (4):71-73.

Buttel, F.H. (1987b). Social science institutions, knowledge, and tools to address problems and issues. *Proceedings of Phase I Workshop*. Neill Schaller, compiler, Social Science Agricultural Agenda Project, Spring Hill Conference Center, Minneapolis, MN, June 9-11, 1987. 79-115.

Buttel, F.H. & Belsky, J. (1987). Biotechnology, plant breeding, and intellectual property: social and ethical dimensions. *Science, Technology, & Human Values*. 12:31-49.

Buttel, F.H. & Gertler, M.E. (1982). Agricultural structure, agricultural policy and environmental quality: some observations on the context of agricultural research in North America. *Agriculture and Environment*. 7:101-119.

Buttel, F.H. & Newby, H., eds. (1980). *The Rural Sociology of the Advanced Societies*. Allanheld, Osmun, Totowa, NJ.

Byerlee, D., Collinson, M.P., Perrin, R.K., Winkelmann, D.L., Biggs, S., Moscardi, E.R., Martinez, J.C., Harrington, L., & Benjamin, A. (1980). *Planning Technologies Appropriate to Farmers: Concepts and Procedures*. CIMMYT, El Batan, Mexico.

Byrne, E.F. & Pitt, J.C., eds. (1989). *Technological Transformation: Contextual and Conceptual Implications*. Kluwer Academic Publishers, The Netherlands.

Callicott, J.B. (1980). Animal liberation: a triangular affair. *Environmental Ethics*. 2 (4):311-338.

Callicott, J.B. (1986). The search for an environmental ethic. *Matters of Life and Death: New Introductory Essays in Moral Philosophy*. (2nd ed.) (Regan, T., ed.) Random House, New York. 381-424.

Callicott, J.B. (1987). *Companion to A Sand County Almanac*. University of Wisconsin Press, Madison, WI.

Callicott, J.B. (1988). Agroecology in context. *Journal of Agricultural Ethics*. 1 (1):3-9.

Caplan, A.L. (1986). The ethics of uncertainty: the regulation of food safety in the United States. *Agriculture and Human Values*. 3 (1 & 2):180-190.

Cardoso, F.H. & Faletto, E. (1979). *Dependency and Development in Latin America*. (Urquidi, M.M., trans.) University of California Press, Berkeley.

Carson, R. (1962). *Silent Spring*. Houghton Mifflin Company, Boston.

Carter, H., Cochrane, W., Day, L., Powers, R., & Tweeten, L. (1981). *Research and the Family Farm*. Agricultural Experiment Station Committee on Organization and Policy, Cornell University, New York.

Castle, E.N. (1970). Priorities in agricultural economics for the 1970's. *American Journal of Agricultural Economics*. 52:831-840

Cochrane, W.W. (1979). *The Development of American Agriculture: An Historical Analysis*. University of Minnesota Press, Minneapolis, MN.

Collins, J. & Lappe, F.M. (1977). *Food First*. Ballantine Books, New York.

Commoner, B. (1971). *The Closing Circle: Nature, Man and Technology*. Alfred A. Knopf, New York.

Comstock, G., ed. (1987). *Is There a Moral Obligation to Save the Family Farm?* Iowa State University Press, Ames, IA.

Comstock, G. (1988). The case against bGH. *Agriculture and Human Values*. 5 (3):367-352.

Conway, G.R. (1985). Agricultural ecology and farming systems research. *Agricultural Systems Research for Developing Countries: ACIAR Proceedings No. 11.* (Remenyi, J.V., ed.) 43-59.

Council on Environmental Quality and the Department of State. 1981. *The Global 2000 Report to the President: Entering the Twenty-first Century: A Report.* (Barney, G.O., Dir.) The Council, Washington, DC.

Coye, M.J. (1986). Health effects of agricultural production. *New Directions for Agriculture and Agricultural Research.* (Dahlberg, K., ed.) Totowa, NJ. 147-162.

Crocker, D. (1990). The hope for just participatory ecodevelopment in Costa Rica. *Ethics of Conservation and Development: Global Challenge and International Response.* (Engel, R. & Engel J., eds.) University of Arizona Press, Tuscon, AZ. 150-163.

Crosby, Jr., A.W. (1972). *The Columbian Exchange: Biological and Cultural Consequences of 1492.* Greenwood Press, Westport, CT.

Crowder, L.V. (1987). Agents, vendors and farmers: public and private sector extension in agricultural development. *Agriculture and Human Values.* 4(4):26-31.

Dahlberg, K.A. (1979). *Beyond the Green Revolution: The Ecology and Politics of Global Agricultural Development.* Plenum Press, New York.

Dahlberg, K.A. (1988). Ethical and value issues in international agricultural research. *Agriculture and Human Values.* 5 (1 & 2):101-111.

Dahlberg, K.A., ed. (1986). *New Directions for Agriculture and Agricultural Research: Neglected Dimensions and Emerging Alternatives.* Rowman and Allanheld, Totowa, NJ.

Dale, C. (1981). Agricultural research as state intervention. *Science and Agricultural Development.* (Busch, L., ed.) Allanheld, Osmun, Totowa, NJ.

Dalrymple, D.G. (1977). Evaluating the impact of international research on wheat and rice production in the developing nations. *Resource Allocation and Productivity in National and International Agricultural Research.* (Arndt, T.M., Dalrymple, D.G., & Ruttan, V.W., eds.) University of Minnesota Press, Minneapolis, MN. 171-208.

Danbom, D.B. (1986). Publicly sponsored agricultural research in the United States from an historical perspective. *New Directions for Agriculture and Agricultural Research: Neglected Dimensions and Emerging Alternatives.* (Dahlberg, K.A., ed.) Rowman and Allanheld, Totowa, NJ. 142-162.

Danhof, C.H. (1969). *Change in Agriculture: The Northern United States, 1820-1870.* Harvard University Press, Cambridge, MA.

Davis, D.E. (1989). *Ecophilosophy.* R. & E. Miles, San Pedro, CA.

de Castro, J. (1977). *The Geopolitics of Hunger.* The Monthly Review Press, New York.

de Janvry, A. (1977). Inducement of technological and institutional innovations: an interpretive framework. *Resource Allocation and Productivity in National and International Agricultural Research.* (Arndt, T.M., Dalrymple, D.G., & Ruttan, V.W., eds.) University of Minnesota Press, Minneapolis, MN. 551-563.

de Janvry, A. (1982). *Historical Forces that Have Shaped World Agriculture: A*

Structuralist Perspective. Giannini Foundation of Agricultural Economics, University of California at Berkeley, Berkeley, CA.

de Janvry, A. & Dethier, J.J. (1985). *Technological Innovation in Agriculture: The Political Economy of Its Rate and Bias.* CGIAR Study Paper Number 1. The World Bank, Washington, DC.

de Janvry, A. & Garcia, R. (1989). *Rural Poverty and Environmental Degradation in Latin America: Causes, Effects, and Alternative Solutions.* International Fund for Agricultural Development, Rome.

de Janvry, A. & LeVeen, P.E. (1986). Historical forces that have shaped world agriculture: a structural perspective. *New Directions for Agriculture and Agricultural Research: Neglected Dimensions and Emerging Alternatives.* (Dahlberg, K.A., ed.) Rowman and Allanheld, Totowa, NJ. 82-104.

de Janvry, A., LeVeen, P.E., & Runsten, D. (1980). *Mechanization in California Agriculture: The Case of Canning Tomatoes.* IICA, Costa Rica.

de Janvry, A. & Vandeman, A. (1982). The macrocontext of rural development: lessons from the United States experience. Paper presented at the Conference on US-Mexico Agriculture and Rural Development. Mexico City.

Devall, B. & Sessions, G. (1984). The development of natural resources and the integrity of nature. *Environmental Ethics.* 6 (4):293-322.

DeWalt, B.R. (1985). Anthropology, sociology, and farming systems research. *Human Organization.* 44:106-114.

DeWalt, B.R. (1988a). Halfway there: social science in agricultural development and the social science of agricultural development. *Human Organization.* 47:343-53.

DeWalt, B.R. (1988b). The cultural ecology of development: ten precepts for survival. *Agriculture and Human Values.* 5 (1 and 2):112-23.

DeWalt, B.R. & Hudgens, R. (1988). *Farming Systems Research and Extension Activities in Guatemala: A Results Inventory.* Office of Arid Lands Studies, University of Arizona, Tucson, AZ.

DeWalt, K.M. & DeWalt, B.R. (1989). Incorporating nutrition into agricultural research: a case study from Southern Honduras. *Making Our Research Useful: Case Studies in the Utilization of Anthropological Knowledge.* (Rylko-Bauer, B., McElroy, A. & Willigen, J. van, eds.) Westview Press, Boulder, CO.

DeWalt, K.M., DeWalt, B.R., Escudero, J.C., & Barkin, D. (In production). The nutrition effects of shifts from maize to sorghum production in four Mexican communities. *Food Policy.*

Dewey, K.G. (1981). Nutritional consequences of the transformation from subsistence to commercial agriculture in Tabasco, Mexico. *Human Ecology.* 9:151-188.

Diamond, J. (1987). The worst mistake in the history of the human race. *Discover.* May:64-6.

Diesing, P. (1982). *Science & Ideology in the Policy Sciences.* Aldine, Hawthorne, NY.

Dorner, P. (1983). Technology and U.S. agriculture. *Technology and Social Change in Rural Areas.* (Summers, G., ed.) 73-86. Westview Press, Boulder, CO.

Douglass, G., ed. (1984). *Agricultural Sustainability in a Changing World Order.* Westview Press, Boulder, CO.

Doyle, J. (1985). *Altered Harvest.* Viking Penguin, New York.

Drengson, A. (1982). Four philosophies of technology. *Philosophy Today* (Summer).

Dundon, S. (1983). Hidden obstacles to creativity in agricultural science. *Proceedings of the Agriculture, Change, and Human Values Conference.* (Haynes, R., ed.) University of Florida, Gainesville, FL. 2:836-868.

Dundon, S. (1986). The moral factor in innovative research. *The Agricultural Scientific Enterprise: A System in Transition.* (Busch, L. & Lacy, W.B., ed.) 39-51. Westview Press, Boulder, CO.

Easterbrook, G. (1985). Making sense of agriculture: a revisionist look at farm policy. *The Atlantic.* July:63-78. reprinted in *Is There a Moral Obligation to Save the Family Farm?* (Comstock, G., ed.) Iowa State University Press, Ames, IA.

Ebenreck, S. (1983). A partnership farmland ethic. *Environmental Ethics.* 5 (1):33-45.

Ebenreck, S. (1984). Stopping the raid on soil: ethical reflections on "Sodbusting" legislation. *Agriculture and Human Values.* 1 (3):3-9.

Eberstad, N. (1980). Malthusians, marxists and missionaries. *Society.* 17 (September/October):29-35.

Economic Research Service. (1987). Economic indicators of farm sector costs of production. USDA-ERS, ECIFS 7-3.

Eicher, C.K. (1984). Facing up to Africa's food crisis. *Foreign Affairs.* 61 (1):154-174. reprinted in *Agricultural Development in the Third World.* (Eicher, C.K. & Staatz, J.M., eds.) 453-479.

Eicher, C.K. & Staatz, J.M., eds. (1984). *Agricultural Development in the Third World.* John Hopkins University Press, Baltimore, MD.

Eisenberg, E. (1989). Back to Eden. *The Atlantic.* November:57-89.

Elton, C.S. (1926). *Animal Ecology.* Sidgwick & Jackson, Ltd., London.

Emerson, R.W. (1870, reprinted 1904). Farming. *Society and Solitude.* Houghton Mifflin, Boston. 139-154.

Engel, R. & Engel, J., eds. (1990). *Ethics of Conservation and Development: Global Challenge and International Response.* University of Arizona Press, Tuscon, AZ.

EPA (1982). *Chesapeake Bay: Introduction to an Ecosystem.* Environmental Protection Agency, Washington, DC.

Esmay, M.L. (1983). Appropriate Agricultural Mechanization for Upland Farming in Thailand. *UNDP/FAO.* Michigan State University, East Lansing, MI.

Esmay, M.L. (1986). *Report on Trip to China and Japan.* Michigan State University, East Lansing, MI.

Esmay, M.L. (1987). *Manpower Needs Study for Agricultural Engineering in Zimbabwe.* Michigan State University, International Institute of Agriculture, East Lansing, MI.

Evenson, R.E. (1983). *Intellectual Property Rights and Agribusiness Research and Development: Implications for the Public Agricultural Research System.* American Agricultural Economics Association.

Evenson, R.E. & Putnam, J.D. (1987). *Institutional Change in Intellectual Property Rights.* American Agricultural Economics Association.

Evenson, R.E., Waggoner, P.E., & Ruttan, V.W. (1979). Economic benefits from research: an example from agriculture. *Science.* 205:1101-1107.

Farrell, K.R. (1989). Contemporary issues: ethics in agricultural research. *Journal of Production Agriculture.* 2 (3):193-196.

Fink, D. (1986). Constructing rural culture: family and land in Iowa. *Agriculture and Human Values.* 3 (4):43-53.

Fite, G. (1981). *American Farmers: The New Minority.* Indiana University Press, Bloomington, IN.

Flader, S. (1974). *Thinking Like a Mountain.* University of Nebraska Press, Lincoln, NB.

Fletcher, J. (1977). Give if it helps but not if it hurts. *World Hunger and Moral Obligation.* (Aiken, W. & La Follette, H., eds.) 103-114.

Fleuret, P. & Fleuret, A. (1980). Nutrition, consumption and agricultural change. *Human Organization.* 39:259-260.

Flora, C.B. (1985). Women and agriculture. *Agriculture and Human Values.* 2 (1):5-12.

Flora, C.B. (1986). Values and the agricultural crisis: differential problems, solutions, and value constraints. *Agriculture and Human Values.* 3 (4):16-23.

Foster, G. (1967). *Tzintzuntzan: Mexican Peasants in a Changing World.* Little, Brown, Boston.

Fox, M.A. (1986). *The Case for Animal Experimentation: An Evolutionary and Ethical Perspective.* University of California Press, Berkeley.

Fox, M.W. (1976). *Between Animal and Man.* Coard, McCann & Geoghegan, New York.

Fox, M.W. (1980). *Returning to Eden: Animal Rights and Human Responsibility.* Viking Press, New York.

Fox, M.W. (1984). *Farm Animals: Husbandry, Behavior, and Veterinary Practice: Viewpoints of a Critic.* University Park Press, Baltimore, MD.

Fox, M.W. (1986). *Agricide: The Hidden Crisis that Affects Us All.* Schocken, New York.

Francis, C.A. (1988). Sustainable versus nonsustainable resources: options for tomorrow's agriculture. *National Forum, The Phi Kappa Phi Journal.* Louisiana State University, Baton Rouge, LA. Summer.

Francis, C.A. & Hildebrand, P.E. (1989). Farming systems research and extension (FSR/E) in support of sustainable agriculture. *Farming Systems Research-Extension Newsletter.* Institute of Food and Agricultural Sciences, University of Florida, Gainesville, FL.

Frank, A.G. (1969). *Latin America: Underdevelopment or Revolution.* Monthly Review Press, New York.

Franke, R.W. (1988). Power, class, and traditional knowledge in Sahel food production. *Studies in Power and Class in Africa.* (Markovitz, I.L., ed.) Oxford University Press, New York. 237-285.

Franke, R.W. & Chasin, B.H. (1980). *Seeds of Famine: Ecological Destruction and the Development Dilemma in the West African Sahel.* Allanheld, Osmun and Co., Montclair, NJ.

Friedland, W.H. & Barton, A.E. (1975). *Destalking the Wily Tomato: A Case Study in Social Consequences in California Agricultural Research.* University of California Davis Department of Applied Behavioral Research, Monograph 2.

Friedland, W.H., Barton A.E., & Thomas, R.J. (1981). *Manufacturing Green Gold: Capital, Labor and Technology in the Lettuce Industry.* Cambridge University Press, Cambridge, MA.

Futrell, G., Dhuynetter, K., & O'Brien, D. Trends in hog production efficiency 1970-1987. *Economic Report Series No. 22 (Revised).* Dept. of Economics, Iowa State University, Ames, IA.

Gajbhiye, H. & Hadwiger, D.F. (1987). Political support for national and international public research. *Public Policy and Agricultural Research.* (Hadwiger, D.F. & Browne, W.P., eds.) St. Martin's Press, New York.

Gardner, D. & Pope, R. (1978). *How is scale and structure determined in agriculture?* American Journal of Agricultural Economics. 60:235-302.

Gendron, B. (1977). *Technology and the Human Condition.* St. Martin's Press, New York.

George, S. (1977). *How the Other Half Dies: The Real Reasons for World Hunger.* Allanheld, Osmun and Co., Montclair, NJ.

George, S. (1984). *Ill Fares the Land: Essays on Food, Hunger and Power.* Institute for Policy Studies, Washington, DC.

Gillespie, G.W. & Buttel, F.H. (1989). Farmer ambivalence toward agricultural research: an empirical assessment. *Rural Sociology.* 54 (3):382-408.

Gladwin, C.H. (1985). Values and goals of Florida farm women: do they help the family farm survive? *Agriculture and Human Values.* 2 (1):40-47.

Gladwin, C.H. (1989). Indigenous knowledge systems, the cognitive revolution, and agricultural decision making. *Agriculture and Human Values.* 4 (3):32-41.

Glass, B. (1955). Genetics in the Service of Man. *Johns Hopkins Magazine.* 6 (5):2-8.

Gliessman, S.R. (1990). *Agroecology: Researching the Ecological Basis for Sustainable Agriculture.* Springer-Verlag, New York.

Goldschmidt, W. (1947, republished 1978). *As You Sow: Three Studies in the Social Consequences of Agribusiness.* Allanheld, Osmun and Co., Montclair, NJ.

Goldstein, D.J. (1988). Molecular biology and the protection of germplasm: a matter of national security. *Seeds and Sovereignty: The Use and Control of Plant Genetic Resources.* (Kloppenburg, J.R., Jr., ed.) 315-337. Duke University Press, Durham, NC.

Goodwyn, L. (1978). *The Populist Moment.* Oxford University Press, New York.

Goulet, D. (1973). *The Cruel Choice.* Atheneum, NY.

Grabowski, R. (1979). The implications of an induced innovation model. *Economic Development and Cultural Change.* 27 (July):723-734.

Graham, T.E., ed. (1986). *The Agricultural Social Gospel in America: The Gospel of the Farm by Jenkin Lloyd Jones.* Edwin Mellen Press, Lewiston, NY.

Griffin, K. (1972). *The Green Revolution: An Economic Analysis.* United Nations Research Institute for Social Development Research Report No. 72.6, Geneva.

Griffin, K. (1974). *The Political Economy of Agrarian Change: An Essay on the Green Revolution.* Harvard University, Cambridge, MA.

Habermas, J. (1972). *Knowledge and Human Interests.* Beacon, Boston.

Hadwiger, D.F. (1982). *The Politics of Agricultural Research.* University of Nebraska Press, Lincoln, NB.

Hadwiger, D.F. & Browne, W.P., eds. (1987). *Public Policy and Agricultural Technology.* St. Martin's Press, New York.

Hager, P. (1987). UC ordered to alter focus of farm research. *The Los Angeles Times.* November 18, 1987:1A, 22A.

Hannay, N.B. & McGinn, R.E. (1980). The anatomy of modern technology: prolegomenon to an improved public policy for the social management of technology. *Daedalus.* 109 (Winter):25-53.

Hansen, M., Busch, L., & Burkhardt, J. (1986). Plant breeding and biotechnology. *BioScience.* 36:29-39.

Hardin, R. (1982). *Collective Action.* Johns Hopkins University Press, Baltimore, MD. 82, 83.

Hardin, C.M. (1955). *Freedom in Agricultural Education.* University of Chicago Press, New York. 56-61.

Hardin, C.M. (1979). Feeding the world: conflicting view on policy. *Agricultural History.* 53 (October):787-795.

Hardin, G. (1974). Living on a lifeboat. *BioScience.* 24:561-568. reprinted in *Stalking the Wild Taboo.* (2nd ed.) 221-241.

Hardin, G. (1976a). The cybernetics of competition. *Perspectives in Biology and Medicine.* 7:58-84. reprinted in *Stalking the Wild Taboo.* (2nd ed.) 148-176.

Hardin, G. (1976b). Carrying capacity as an ethical concept. *Soundings.* 59(1):120-137. reprinted in *Stalking the Wild Taboo.* (2nd ed.) 245-261.

Hargrove, E.C. (1987). The value of environmental ethics. *The Environmental Professional.* 9 (4):289-294.

Hargrove, E.C. (1989). *Foundations of Environmental Ethics.* Prentice-Hall, Englewood Cliffs, NJ.

Harlan, J.R. (1988). Seeds and sovereignty: an epilogue. *Seeds and Sovereignty: The Use and Control of Plant Genetic Resources.* (Kloppenburg, J.R., Jr., ed.) 356-362. Duke University Press, Durham, NC.

Harris, C.E., Jr. (1986). *Applying Moral Theories.* Wadsworth, Belmont, CA.

Harris, L.D. (1984). *The Fragmented Forest: Island Biogeography Theory and the Preservation of Biotic Diversity.* University of Chicago Press, Chicago.

Harrison, R. (1964). *Animal Machines: The New Factory Farming Industry.* V. Steward, London.

Hassebrook, C. & Hegyes, G. (1988). *Choices for the Heartland: Alternative Directions in Biotechnology and Implications for Family Farming, Rural Communities and the Environment.* Center for Rural Affairs, Walthill, NB.

Harvey, D. (1989). *The Condition of Post-Modernity.* Basil Blackwell, Cambridge, MA.

Hayami, Y. (1984). Assessment of the green revolution. *Agricultural Development in the Third World.* (Eicher, C.K. & Staatz, J.M., eds.) 389-396. Originally published as: Induced innovation, green revolution, and income distribution: comment. *Economic Development and Cultural Change.* 30 (1):169-176, 1981.

Hayami, Y. & Ruttan, V.W. (1985). *Agricultural Development: An International Perspective.* Johns Hopkins University Press, Baltimore, MD.

Hayek, F. A. (1978). The atavism of social justice. *New Studies in Philosophy, Politics, Economics and the History of Ideas.* Routledge and Kegan Paul, London. 58:57-70.

Haynes, R. & Lanier, R., eds. (1982). *Agriculture, Change and Human Values: Proceedings of a Multidisciplinary Conference.* Humanities and Agriculture Program, The University of Florida, Gainesville, FL.

Hays, S.P. (1959). *The Gospel of Efficiency: The Progressive Conservation Movement, 1890-1920.* Harvard University Press, Cambridge, MA.

Hayter, E.W. (1968). *The Troubled Farmer 1850-1900: Rural Adjustment to Industrialism.* Northern Illinois University Press, DeKalb, IL.

Healy, R.G. (1985). Land, agriculture, and the long view. *Agriculture and Human Values.* 2 (4):36-40.

Heffernan, W.D. (1986). Review and evaluation of social externalities. *New Directions for Agriculture and Agricultural Research: Neglected Dimensions and Emerging Alternatives.* (Dahlberg, K., ed.) Rowman and Allanheld, Totowa, NJ. 199-220.

Hewitt de Alcantara, C. (1976). *Modernizing Mexican Agriculture.* United Nations Research Institute for Social Development, Geneva.

Hightower, J. (1973). *Hard Tomatoes, Hard Times.* Schenkman, Cambridge, MA.

Hightower, J. (1975). The case for the family farm. *Food for People, Not for Profit.* (Lerza, C. & Jacobson, M., eds.) Ballantine, New York.

Hightower, J. (1976). *Eat Your Heart Out.* Crown Books, New York.

Hildebrand, P.E. (1980). Motivating small farmers, scientists and technicians to accept change. *Agricultural Administration.* 8 (5):375-383.

Hildebrand, P.E., ed. (1986). *Perspectives on Farming Systems Research and Extension.* Lynne Rienner Publishers, Boulder, CO.

Hirschman, A.O. (1981). *Essays in Trespassing: Economics to Politics and Beyond.* Cambridge University Press, New York.

Hirschman, A.O. (1984). *Getting Ahead Collectively: Grassroots Experiences in Latin America.* Pergamon Press, New York.

Hoag, R. (1986). Happiness and freedom: recent work on John Stuart Mill. *Philosophy and Public Affairs.* 15:188-199.

Holdeman, Q.L. (1983). The scripted good Samaritan. *Transactional Analysis Journal.* 249-250. Hollander, R.D. (1986). Values and making decisions about agricultural research. *Agriculture and Human Values.* 3 (3):33-40.

Horowitz, I.L. (1972). *Three Worlds of Development: The Theory and Practice of International Stratification.* (2nd ed.) Oxford University Press, London.

Horwith, B.J., Windle, P.N., MacDonald, E.F., Parker, J.K., Ruby, A.M., & Elfring, C. (1989). The role of technology in enhancing low-resource agriculture in Africa. *Agriculture and Human Values.* 6 (3):68-84.

Innis, G. (1974). Stability, sensitivity, resilience, persistence. What is of interest? *Ecosystem Analysis and Prediction.* (Levin, S., ed.) Society for Industrial and Applied Mathematics, Philadelphia.

Ishikawa, S. (1967). *Economic Development in Asian Perspective.* Kinokuniya Bookstore Co., Tokyo.

Jackson, W. (1980). *New Roots for Agriculture.* Friends of the Earth, San Francisco.

Jackson, W. (1985). *New Roots for Agriculture.* (2nd ed.) University of Nebraska Press, Lincoln, NB.

Jackson, W. (1987). *Altars of Unhewn Stone: Science and the Earth.* North Point Press, San Francisco, CA.

Jackson, W., Berry, W., and Colman, B., eds. (1984). *Meeting the Expectations of the Land: Essays in Sustainable Agriculture and Stewardship.* North Point Press, San Francisco.

Jacobs, J. (1984). *Cities and the Wealth of Nations.* Random House, New York.

Jefferson, T. (1984). *Writings.* Literary Classics of the United States, New York.

Jennings, B. (1988). *Foundations of International Agricultural Research: Science and Politics in Mexican Agriculture.* Westview Press, Boulder, CO.

Jensen, J.M. (1985). The role of farm women in American history: Areas for Additional Research. *Agriculture and Human Values.* 2 (1):13-17.

Johnson, G.L. (1984). Academia needs a new covenant for serving agriculture. Mississippi State Agricultural and Forestry Experiment Station Special Publication.

Johnson, G.L. (1987). *Research Methodology for Economists.* Macmillan, New York.

Johnson, G.L. & Brown, J.L. (1980). Literature and philosophic ideas underlying model validation and verification. *An Evaluation of the Normative and Prescriptive Content of the Department of Energy Mid-Term Energy Forecasting System (MEFS) and the Texas National Energy Modeling Project (TNEMP).* Energy and National Resources Advisory Council, Austin, TX.

Johnson, R.L. & Johnson, G.V., eds. (1990). *Economic Valuation of Natural Resources: Issues, Theory, and Applications.* Westview Press, Boulder, CO.

Jordan, J.P., O'Connell, P.F., & Robinson, R.R. (1986). Historical evolution of the state agricultural experiment station system. *New Directions for Agriculture and Agricultural Research.* (Dahlberg, K.A., ed.) Rowman and Allanheld, Towota, NJ. 147-162.

Just, R.E., Schmitz, A., & Zilberman, D. (1979). Technological change in agriculture: the sources, diffusion, and social imports of technological change in agriculture are examined. *Science.* 206:1277-1280.

Kalter, R. (1985). The new biotech agriculture: unforeseen economic consequences. *Issues in Science and Technology.* 2:125-133.

Kelling, K.A. (1989). *Proceedings of the University of Wisconsin Sustainable Agriculture "Listening" Meetings.* University of Wisconsin-Cooperative Extension Service, Madison, WI.

Kellogg, C.E. & Knapp, D.C. (1966). *The College of Agriculture: Science in the Public Service.* McGraw-Hill, NY.

Kennan, G.F. (1985). Morality and foreign policy. *Foreign Affairs.* 64 (2):205-218.

Kenney, M. (1986). *Biotechnology: The University-Industrial Complex.* Yale University Press, New Haven, CT.

Khan, A.U. (1979). Agricultural mechanization and farm machinery production in Pakistan. *Monographs on Appropriate Industrial Technology, No. 4.* United Nations Industrial Development Organization, Vienna, Austria.

Khan, A.U. (1985). *The Role of Agricultural Engineers in Development.* American Society of Agricultural Engineers, St. Joseph, MI. ASAE Paper No. 85-5040.

Kirk, D. & Eliason, E.K., eds. (1982). *Food and People.* Boyd & Fraser, San Francisco.

Kirkendall, R.S. (1984). The central theme of American agricultural history. *Agriculture and Human Values.* 1 (2):6-8.

Kirkendall, R.S. (1986). The Agricultural Colleges: Between Tradition and Modernization. *Agricultural History.* 60 (2):3-21.

Kirkendall, R.S. (1987a). A history of the family farm. *Is There a Moral Obligation to Save the Family Farm?* (Comstock, G., ed.) Iowa State University Press, Ames, IA.

Kirkendall, R.S. (1987b). Up to now: a history of American agriculture from Jefferson to revolution to crisis. *Agriculture and Human Values.* 4 (1):4-26.

Kloppenburg, J.R., Jr. (1988a). *First the Seed: The Political Economy of Plant Technology, 1492-2000.* Cambridge University Press, Cambridge, MA.

Kloppenburg, J.R., Jr., ed. (1988b). *Seeds and Sovereignty: The Use and Control of Plant Genetic Resources.* Duke University Press, Durham, NC.

Kloppenburg, J.R., Jr. & Kleinman, D.L. (1988). Seeds of controversy: national property versus common heritage. *Seeds and Sovereignty: The Use and Control of Plant Genetic Resources.* (Kloppenburg, J.R., Jr., ed.) Duke University Press, Durham, NC. 173-203.

Knezek, B.D., O.B. Hesterman, and L. Wink. (1988). Exploring a new vision of agriculture. *National forum, The Phi Kappa Phi Journal.* Louisiana State University, Baton Rouge, LA. Summer.

Knight, F. (1935). *The Ethics of Competition.* Allen & Unwin, London.

Knorr, D., ed. (1986). *Food Biotechnology.* Dekker, New York.

Koestler, A. (1968). *The Ghost in the Machine.* Macmillan, New York.

Kuhn, T.S. (1962). *The Structure of Scientific Revolutions.* University of Chicago Press, Chicago.

Kunkel, H. O. (1984). Agricultural ethics—the setting. *Agriculture and Human Values.* 1 (1):20-23.

Lappe, F.M. (1985). The family farm: caught in the contradictions of American values. *Agriculture and Human Values.* 2 (2):36-43.

Lappe, F.M. & Callicott, J.B. (1989) Marx meets Muir: toward a synthesis of the progressive political and ecological visions. *Global Perspectives on Agroecology and Sustainable Agricultural Systems.* (Allen, P. & Van Dusen, P., eds.) University of California Press, Santa Cruz, CA. 21-30.

Lappe, F.M. & Collins, J. (with C. Fowler). (1977). *Food First: Beyond the Myth of Scarcity.* Houghton Mifflin, Boston.

Lappe, F.M., Collins, J., & Kinley, D. (1980). *Aid as Obstacle: Twenty Questions about Our Foreign Aid and the Hungry.* Institute for Food and Development Policy, San Francisco.

Latour, B. (1987). *Science in Action.* Harvard University Press, Cambridge, MA.

Lele, U. (1981). Rural Africa: modernization, equity, and long-term development. *Science.* 211 (Feb. 6):547-553. reprinted in *Agricultural Development in the Third World.* (Eicher, C.K. & Staatz, J.M., eds.) 436-452.

Lemons, J. (1986). Structural trends in agriculture and preservation of family farms. *Environmental Management.* 10 (1):75-88.

Leontief, W., Carter, A., & Petrin, P. (1977). *The Future of the World Economy: A United Nations Study.* Oxford University Press, New York.

Leopold, A. (1939). A biotic view of land. *Journal of Forestry.* 37:727.

Leopold, A. (1949). *A Sand County Almanac.* Oxford University Press, London.

Leopold, A. (1979). Some fundamentals of conservation in the Southwest. *Environmental Ethics.* 1:136-137.

Lerner, D. (1958). *The Passing of Traditional Society: Modernizing the Middle East.* Free Press, New York.

Lerza, C., & Jacobson, M., eds. (1975). *Food for People, Not for Profit.* Ballentine, New York.

Levin, S., ed. (1974). *Ecosystem Analysis and Prediction.* Society for Industrial and Applied Mathematics, Philadelphia.

Levine, J. (1986). Hearts and minds: the politics of diet and heart disease. *Consuming Fears.* (Sapolsky, H.M., ed.) Basic Books, New York.

Levins, R. (1973). Fundamental and Applied Research in Agriculture. *Science.* 181:523-524.

Levins, R. & Lewontin, R. (1985). *The Dialectical Biologist.* Harvard Univ. Press, Boston.

Lewontin, R. (1969). The meaning of stability. *Diversity and Stability in Ecological Systems.* (Woodwell, G.M. & Smith, H.H., eds.) Brookhaven National Laboratory Publication No. 22, Clearinghouse for Federal Scientific and Technical Information, Springfield, VA.

Liebman, E. (1983). *California Farmland: A History of Large Agricultural Landholdings.* Rowman and Allanheld, Totowa, NJ.

Lipton, M. & Longhurst, R. (1989). *New Seeds and Poor People.* Johns Hopkins University Press, Baltimore, MD.

Locke, J. (1690, reprinted 1980). *The Second Treatise of Government.* (Macpherson, C.B., ed.) Hackett Publishing Co., Indianapolis, IN.

Lockeretz, W. (1988). Issues in sustainable agriculture. Prepared for the Rural Economic Policy Program. Aspen Institute, Washington, DC.

Lockeretz., W., ed. (1983). *Environmentally Sound Agriculture.* Praeger, New York.

Logsdon, G. (1984). The importance of traditions farming practices for a sustainable modern agriculture. *Meeting the Expectations of the Land: Essays in Sustainable Agriculture and Stewardship.* (Jackson, W., Berry, W., & Colman, B., eds.) North Point Press, San Francisco.

Lowrance, W.W. (1986). *Modern Science and Human Values.* Oxford University Press, New York.

MacIntyre, A. (1981). *After Virtue.* University of Notre Dame Press, Notre Dame, IN.

MacIntyre, A. (1988). *Whose Justice? Which Rationality?* University of Notre Dame Press, Notre Dame, IN.

MacNeill, J. (1989). Strategies for sustainable economic development. *Scientific American* Sept.:155-174.

Madden, P. (1986). Beyond conventional economics—an examination of the values implicit in the neoclassical economic paradigm as applied to the evaluation of agricultural research. *New Directions for Agriculture and Agricultural Research: Neglected Dimensions and Emerging Alternatives.* (Dahlberg, K.A., ed.) 221-258. Rowman and Allanheld, Totowa, NJ.

Madden, P. (1987). Toward a new covenant of agricultural academe. *Public Policy and Agricultural Technology.* (Hadwiger, D.F. & Browne, W.P., eds.) St. Martin's Press, New York.

Madden, P. (1989). Policy Options for a More Sustainable Agriculture. *Increasing Understanding of Public Problems and Policies.* Farm Foundation, Oak Brook IL.

Madden, P. & Brewster, D.E., eds. (1970). *A Philosopher Among Economists: Selected Works of John M. Brewster.* J. T. Murphy, Philadelphia.

Madden, P. & Thompson, P.B. (1987). Ethical perspectives on changing agricultural technology in the U.S. *Journal of Law, Ethics and Public Policy.* 3 (1):85-116.

Marcus, A.I. (1985). *Agricultural Science and the Quest for Legitimacy.* Iowa State University Press, Ames, IA.

Marcus, A.I. (1987). Constituents and constituencies: an overview of the history of public agricultural research institutions in America. *Public Policy and Agricultural Research.* (Hadwiger, D.F. & Browne, W.P., eds.) St. Martin's Press, New York.

Margalef, R. (1968). *Perspectives in Ecological Theory.* University of Chicago Press, Chicago.

Markovitz, L.L., ed. (1988). *Studies in Power and Class in Africa.* Oxford University Press, New York.

Marshall, E. (1980). Bergland Opposed on Farm Machinery Policy. *Science.* 208:578-580.

Martin, P. & Olmstead, A. (1985). The agricultural mechanization controversy. *Science.* 227:601-606.

Massey, D.S. (1988). Economic development and international migration. *Population and Development Review.* 14:383-413.

Maxwell, S. & Fernando, A. (1989). Cash crops in developing countries: the issues, the facts, the policies. *World Development.* 17:1677-1708.

McClelland, D. (1961). *The Achieving Society.* Van Nostrand, Princeton, NJ.

McClennan, E.F. (1983). Rational choice and public policy: a critical survey. *Social Theory and Practice.* 9 (2-3):335-379.

McCorkle, C. (1989). Toward a knowledge of local knowledge and its importance for agricultural RD&E. *Agriculture and Human Values.* 4 (3):4-12.

McKibben, B. (1990). Prophet in Kentucky. *The New York Review of Books.* 37 (10):30-34.

Meeks, M.D. (1985). God and land. *Agriculture and Human Values.* 4 (2):16-27.

Meine, C. (1988). *Aldo Leopold: His Life and Work.* University of Wisconsin Press, Madison, WI.

Merrill, R. (1976). *Radical Agriculture.* Harper, Colophon, New York.

Mill, J.S. (1859, republished 1978). *On Liberty.* (Rapaport, E., ed.) Hackett Publishing Co., Indianapolis, IN.

Mill, J.S. (1861, republished 1979). *Utilitarianism.* (Sher, G., ed.) Hackett Publishing Co., Indianapolis, IN.

Mohammadi, S.B. (1981). American capitalism and agricultural development. *Science and Agricultural Development.* (Busch, L., ed.) Allanheld, Osmun, Totowa, NJ.

Mohr, H. (1979). The ethics of science. *Interdisciplinary Science Reviews.* 4 (1):45-53.

Molnar, J.J. & Beaulieu, L.J. (1984). Societal implications of changes in the organization of agricultural production. *Agriculture and Human Values.* 1 (4):38-44.

Montmarquet, J.A. (1985). Philosophical foundations for agrarianism. *Agriculture and Human Values.* 2 (2):5-14.

Montmarquet, J.A. (1989). *The Idea of Agrarianism: From Hunter-Gatherer to Agrarian Radical in Western Culture.* University of Idaho Press, Moscow, ID.

Mooney, P.R. (1979). *Seeds of the Earth: A Private or Public Resource?* Inter Pares, Ottawa.

Mortimer, R.C. (1950). *Christian Ethics.* Hutchinson, London.

Naess, A. (1973). The shallow and the deep, long-range ecology movement: a summary. *Inquiry.* 16:95-100.

Naess, A. (1984). A defense of the deep ecology movement. *Environmental Ethics.* 6 (3):265-270.

Nagel, T. (1977). Poverty and food: why charity is not enough. *Food Policy: The Responsibility of the United States in the Life and Death Choices.* (Brown, P.G. & Shue, H., eds.) Free Press, New York. 54-62.

Nagel, T. (1986). *The View from Nowhere.* Oxford University Press, Oxford.

Nash, R.F. (1982). *Wilderness and the American Mind.* (3rd ed.) Yale University Press, New Haven, CT.

Nash, R.F. (1989). *The Rights of Nature: A History of Environmental Ethics.* University of Wisconsin Press, Madison, WI.

National Research Council. (1972). *Report of the Committee on Research Advisory to the U.S. Department of Agriculture.* National Academy of Sciences, Washington, DC.

National Research Council. (1975). *Agricultural Production Research Efficiency.* National Academy of Sciences, Washington, DC.

National Research Council. (1989). *Alternative Agriculture.* National Academy Press, Washington, DC.

Norgaard, R.B. (1984). Coevolutionary Agricultural Development. *Economic Development and Cultural Change.* 60:160-173.

Norman, D.W. (1979). The farming systems approach: relevancy for small Farmers. *Increasing the Productivity of Small Farms.* (Karaspan, A.S., ed.) 133-152.

Norman, D.W. & Collison, M. (1985). Farming systems research in theory and practice. *Agricultural Systems Research for Developing Countries: ACIAR Proceedings No. 11.* (Remenyi, J.V., ed.) 16-30.

Norman, R. (1983). *The Moral Philosophers: An Introduction to Ethics.* Clarendon Press, Oxford.

Norton, B.G. (1982a). Environmental ethics and the rights of future generations. *Environmental Ethics.* 4 (4):319-337.

Norton, B.G. (1982b). Environmental ethics and nonhuman rights. *Environmental Ethics.* 4 (1):17-36.

Norton, B.G. (1985). Agricultural development and environmental policy: conceptual issues. *Agriculture and Human Values.* 2 (1):63-70.

Norton, B.G. (1986). Conservation and preservation: a conceptual rehabilitation. *Environmental Ethics.* 8 (3):195-200.

Norton, B.G. (1987). *Why Preserve Natural Variety?* Princeton University Press, Princeton, NJ.

Norton, B.G. (In production, a). *Convergence.* Oxford University Press. Chapter 2.

Norton, B.G. (In production, b). *The Unity of Environmentalism.* Oxford University Press, New York.

Nozick, R. (1974). *Anarchy State and Utopia.* Basic Books, New York.

O'Neill, O. (1980). The moral perplexities of famine relief. *Matters of Life and Death: New Introductory Essays in Moral Philosophy.* (Regan, T., ed.) Random House, New York.

O'Neill, O. (1986). *Faces of Hunger.* Allen & Unwin, London.

O'Neill, R.V. (1986). *A Hierarchical Concept of Ecosystems.* Princeton University Press, Princeton, NJ.

Oasa, E.K. & Swanson, L.E. (1986). The limits of farming systems research and development: should development administrators be interested? *Agricultural Administration.* 23 (4):201-221.

Office of Technology Assessment. (1981). *An Assessment of the United States Food and Agricultural Research System.* U.S. Government Printing Office, Washington, DC.

Office of Technology Assessment. (1985). Congress of the United States, technology, public policy, and the changing structure of American agriculture: a special report for the 1985 farm bill. Washington, D.C. OTA-F, U.S. Government Printing Office.

Office of Technology Assessment. (1986). *Technology, Public Policy, and the Changing Structure of American Agriculture.* Government Printing Office, Washington, DC.

Office of Technology Assessment. (1988). *Enhancing Agriculture in Africa: A Role for U.S. Development Assistance.* OTA-F-356. U.S. Government Printing Office, Washington, DC.

Paarlberg, D. (1981). The land grant colleges and the structure issue. *American Journal of Agricultural Economics.* 63 (1):129-34.

Paddock, W. & Paddock, P. (1967). *Famine, 1975! America's Decision: Who Will Survive?* Little, Brown, Boston.

Page, T. (1977). *Conservation and Economic Efficiency.* Resources for the Future, Washington, DC.

Parker, R.C. & Connor, J.M. (1987). Estimates of Consumer Loss Due to Monopoly in the U.S. Food-Manufacturing Industries. *American Journal of Agricultural Economics.* 61 (November 1979):626-39.

Partridge, E., ed. (1981). *Responsibilities to Future Generations.* Prometheus Books, Buffalo, NY.

Pasour, E.C., Jr. (1973). Economic growth and agriculture: an evaluation of the compensation principle. *American Journal of Agricultural Economics.* 55 (4, pt. 1):611-616.

Passmore, J. (1974). *Man's Responsibility for Nature: Ecological Problems and Western Traditions.* Scribner, New York.

Pattee, H.H. (1973). The physical basis and origin of hierarchical control. *Hierarchy Theory.* (Pattee, H.H., ed.) Braziller, New York.

Paul, C. & DeWalt, B.R., eds. (1985). *El Sorgo en Sistemas de Producción en América Latina.* INTSORMIL/ICRISAT/CIMMYT, Mexico.

Pearce, D., Barbier, E., & Markandya, A. (1987). Sustainable development and cost-benefit analysis. London Environmental Economics Center, Paper 88-01.

Pearse, A. (1980). *Seeds of Plenty, Seeds of Want: Social and Economic Implications of the Green Revolution.* Oxford University Press, New York.

Penn, J. (1979). The structure of American agriculture. *Structure Issues in American Agriculture.* USDA, Washington, DC.

Perelman, M. (1977). *Farming for Profit in a Hungry World: Capital and the Crisis in Agriculture.* Allanheld, Osmun, Towota, NJ.

Perkins, J.H. (1982). *Insects, Experts and The Insecticide Crisis: The Quest for New Pest Management Strategies.* Plenum Press, New York.

Peterson, E.W.F. (1986). Agricultural structure and economic adjustment. *Agriculture and Human Values.* 3 (4):6-15.

Peterson, R. W. (1988). Education for the professional generalist. Letter & Science Magazines. University of Wisconsin-Madison, Spring. 17-21.

Petulla, J.M. (1980). *American Environmentalism: Values, Tactics, and Priorities.* Texas A&M University Press, College Station, TX. 16:204-205.

Petulla, J.M. (1989). The object of environmental ethics. *Environmental Management.* 13 (3):273-278.

Pimentel, D. & Hall, C.W., ed. (1989). *Food and Natural Resources.* Academic Press, Inc., San Diego, CA.

Pimentel, D. & Pimentel, S. (1986). Energy and other natural resources used by agriculture and society. *New Directions for Agriculture and Agricultural Research: Neglected Dimensions and Emerging Alternatives.* (Dahlberg, K., ed.) Rowman and Allanheld, Totowa, NJ. 259-287.

Pinchot, G. (1947). *Breaking New Ground.* Harcourt Brace, New York.

Plunknett, D.L. & Smith, N.J.H. (1982). Agricultural research and third world food production. *Science.* 217:215-20.

Pollock-Byrne, J.M. (1984). *Ethics in Crime and Justice Dilemmas and Decisions.* Brooks/Cole Publishing Company, Pacific Grove, CA.

Rabb, R.L. (1984). Review of *Insects, Experts, and the Insecticide Crisis: The Quest for New Pest Management Strategies* by John Perkins in *Agriculture and Human Values.* 1 (1):34-37.

Randall, A. (1981). *Resource Economics.* Grid Publishing, Columbus, OH.

Randall, A. (1987). *Resource Economics: An Economic Approach to Natural Resource and Environmental Policy.* (2nd Ed.) John Wiley & Son, New York.

Rasmussen, W.D. (1968). Advances in American agriculture: the mechanical tomato harvest as a case study. *Technology and Culture.* 9 (October):531-543.

Ratner, M. (1990). Survey and Opinions; Barriers to Field-Testing Genetically Modified Organisms. *Bio/Technology.* 8 (March):196-198.

Rawlins, S. (1988). Systems science and agricultural sustainability. *National Forum, The Phi Kappa Phi Journal.* Louisiana State University, Baton Rouge, LA. Summer.

Rawls, J. (1971). *A Theory of Justice.* Belnap Press, Cambridge, MA.

Regan, T. (1983). *The Case for Animal Rights.* University of California Press, Berkeley, CA.

Regan, T. (1984). *Earthbound.* Random House, New York.

Regan, T., ed. (1986). *Matters of Life and Death: New Introductory Essays in Moral Philosophy.* Random House, New York.

Remenyi, J.V. (1985). *Agricultural Systems Research for Developing Countries: ACIAR Proceedings No. 11.*

Rescher, N. (1983). *Risk: A Theoretical Introduction to the Theory of Risk Evaluation and Management.* Washington, DC.

Reutlinger, S. & Selowsky, M. (1975). Malnutrition and poverty: magnitude and policy options. *World Bank Occasional Paper, No. 23.* The Johns Hopkins University Press, Baltimore, MD.

Richards, D. (1977). *The Moral Criticism of Law.* Dickenson Publishing, Encino, CA.

Richards, P. (1985). *Indigenous Agricultural Revolution.* Hutchinson, London.

Robbins, L. (1935). *An Essay on the Nature and Significance of Economic Science.* Macmillan, London.

Rodale, R. (1988). Agricultural systems: the importance of sustainability. *National Forum, The Phi Kappa Phi Journal.* Louisiana State University, Baton Rouge, LA. Summer.·

Rogers, E. (1969). *Modernization Among Peasants: The Impact of Communication.* Holt, Rinehart, Winston, New York.

Rogers, E. (1971). *Communication of Innovations: A Cross-Cultural Approach.* Collier-Macmillan, Toronto.

Rollin, B. (1981). *Animal Rights and Human Morality.* Prometheus, Buffalo, NY.

Rollin, B. (1989). *The Unheeded Cry: Animal Consciousness, Animal Pain, and Science.* Oxford University Press, New York.

Rolston, H. (1975). Is there an ecological ethic? *Ethics.* 85 (4):93-109.

Rolston, H. (1988). *Environmental Ethics: Duties to and Values in the Natural World.* Temple University Press, Philadelphia.

Rosenberg, C. (1976). *No Other Gods.* Johns Hopkins University Press, Baltimore, MD.

Ross, P.J. (1985). A commentary on research on American farm women. *Agriculture and Human Values.* 2 (1):19-30.

Rostow, W.W. (1971). *The Stages of Economic Growth.* Cambridge University Press, Cambridge, MA.

Rostow, W.W., ed. (1963). *The Economics of Take-Off into Sustained Growth.* Macmillan, London.

Routley, R. (1973). Is there a need for a new, an environmental ethic? *International Congress of Philosophy 1973, Proceedings.* 15 (1):205-210.

Ruckelshaus, W.D. (1989). Toward a sustainable world. *Scientific American.* Sept.

Rudner, R. (1972). *Philosophy of Social Science.* Prentice-Hall, Englewood Cliffs, NJ.

Ruttan, V.W. (1982a). *Agricultural Research Policy.* University of Minnesota Press, Minneapolis.

Ruttan, V.W. (1982b). Agricultural scientists or reluctant revolutionaries. *Interdisciplinary Science Review.* 7:170-177.

Ruttan, V.W. (1983a). Moral responsibility in agricultural research. *Southern Journal of Agricultural Economics.* 15:73-80.

Ruttan, V.W. (1983b). Accountability in research: examples from agriculture. *Optimum Utilization of Knowledge.* (Senesh, L. & Boulding, K., eds.) Westview Press, Boulder, CO.

Ruttan, V.W. (1986). Toward a global agricultural research system: a personal view. *Research Policy.* 15:307-327.

Ruttan, V.W. (1987a). *Agricultural Research Policy and Development.* FAO Research and Technology Paper 2. Food and Agriculture Organization of the United Nations, Rome.

Ruttan, V.W. (1987b). Commentary: agricultural scientists as reluctant revolutionaries. *Choices.* Third Quarter:3.

Ruttan, V.W. (1988). Cultural endowments and economic development: what can we learn from anthropology? *Economic Development and Cultural Change.* 36 (April supplement):S246-271.

Ruttan, V.W. & Hayami, Y. (1972). Strategies for agricultural development. *Food Research Institute Studies in Agricultural Economics, Trade and Development.* 9 (2):129-148.

Rylko-Bauer, B., McElroy, A., & Willigen, J. van, eds. (1989). *Making Our Research Useful: Case Studies in the Utilization of Anthropological Knowledge.* Westview Press, Boulder, CO.

Sachs, C. (1985). Women's work in the U.S.: variations by region. *Agriculture and Human Values.* 2 (1):31-39.

Sagoff, M. (1988a). On teaching a course on ethics, agriculture, and the environment: part 1. *Journal of Agricultural Ethics.* 1 (2):69-84.

Sagoff, M. (1988b). On teaching a course on ethics, agriculture, and the environment: part 2. *Journal of Agriculture Ethics.* 1 (2):87-100.

Sagoff, M. (1988c). *The Economy of the Earth.* Cambridge University Press, New York.

Sandman, P.M. (1989). Hazard versus outrage: expert versus public perception of food risks. *Intercollegiate Nutrition Consortium*, University of Minnesota, October 31, 1989. St. Paul, MN.

Sapontzis, S.F. (1987). *Morals, Reason, and Animals.* Temple University Press, Philadelphia.

Sauer, R.J. (1987). From the perspective of users and affected persons. *Proceedings of Phase I Workshop.* Niell Schaller, compiler, Social Science Agricultural Agenda Project, Spring Hill Conference Center, Minneapolis, MN, June 9-11, 1987. 267-280.

Schaller, N., compiler. (1987). *Proceedings of Phase I Workshop.* Social Science Agricultural Agenda Project, Spring Hill Conference Center, Minneapolis, MN, June 9-11, 1987.

Schell, O. (1984). *Modern Meat.* Random House, New York.

Scherer, D. & Attig, T., eds. (1983). *Ethics and the Environment.* Prentice-Hall, Englewood Cliffs, NJ.

Schmitz, A., & Seckler, D. (1970). Mechanized agriculture and social welfare: the case of the mechanical tomato harvester. *American Journal of Agricultural Economics.* 52:569-577.

Schooler, D. Jr. (1972). *Science, Scientists and Public Policy.* The Free Press, New York.

Schuh, G.E. (1976). The new macroeconomics of agriculture. *American Journal of Agricultural Economics.* 58:802-811.

Schuh, G.E. (1986). Revitalizing land grant universities. *Choices.* Second Quarter.

Schuh, G.E. & Tollini, H. (1979). *Costs and Benefits of Agricultural Research: The State of the Arts.* World Bank Working Paper 360. World Bank, Washington DC.

Schultz, B. (1985). Sociopolitical implications of stability in agriculture. *Agriculture and Human Values.* 2 (2):60-62.

Schultz, T.W. (1964). *Transforming Traditional Agriculture.* Yale University Press, New Haven, CT.

Schultz, T.W. (1984). The economics of agricultural research. *Agricultural Development in the Third World.* (Eicher, C.K. & Staatz, J.M., eds.) 335-347.

Schumacher, E.F. (1972). *Small Is Beautiful.* Harper & Row, New York.

Schweikhardt, D.B. & Bonnen, J.T. (1986). Policy conflicts in agricultural research: historical perspective and today's challenges. *The Agricultural Scientific Enterprise: A System in Transition.* (Busch, L. & Lacy, W.B., eds.) Westview Press, Boulder, CO. 13-27.

Sen, A. (1981). *Poverty and Famines: An Essay on Entitlement and Deprivation.* Clarendon, Oxford.

Sen, A. (1985). Well-being, agency and freedom, the Dewey lectures 1984. *Journal of Philosophy.* 82 (4):169-221.

Senesh, L. & Boulding, K., eds. (1983). *Optimum Utilization of Knowledge.* Westview Press, Boulder, CO.

Shaner, W.W., Philipp, P.F., & Schmehl, W.R. (1982). *Farming Systems Research and Development: Guidelines for Developing Countries.* Westview Press, Boulder, CO.

Shepard, P. (1984). Moral conflict in agriculture: conquest or moral coevolution? *Agriculture and Human Values.* 1 (4):17-25.

Shepard, P.T. & Hamlin, C. (In production). *Ideology and the Prospect for Consensus in U. S. Agriculture.* Cambridge University Press, New York.

Shepard, P.T. (1989). Impartiality and interpretive intervention in technical controversy. *Technological Transformation: Contextual and Conceptual Implications.* (Byrne, E.F. & Pitt, J.C., eds.) Kluwer Academic Publishers, Chicago. 47-65.

Shils, E. (1974). Faith, utility and the legitimacy of science. *Daedulus.* 103:1-15.

Shover, J. (1976). *First Majority - Last Minority: The Transformation of Rural Life in America.* Northern Illinois University Press, DeKalb, IL.

Shrader-Frechette, K.S. (1988). Agriculture, ethics, and restrictions of property rights. *Journal of Agricultural Ethics.* 1:21-40.

Shue, H. (1980). *Basic Rights.* Princeton University Press, Princeton, NJ.

Shull, G.H. (1946). Hybrid Seed Corn. *Science.* 103 (3 May):547-550.

Shuman, C.B. (1970). Food aid and the free market. *Food Policy.* (Brown, P.G. & Shue, H., eds.) 145-163. The Free Press, New York.

Sinclair, W. (1987). Judge rules research program favors large farms. *The Washington Post.* November 19, 1987:A24.

Singer, P. (1972). Famine, affluence, and morality. *Philosophy and Public Affairs.* 1 (3):229-248.

Singer, P. (1975). *Animal Liberation: A New Ethics for our Treatment of Animals.* Random House, New York.

Singer, P. & Mason, J. (1980). *Animal Factories.* Crown Books, New York.

Smith, V.K. (1990). Can we measure the economic value of environmental amenities? *Southern Economic Journal.* 56:865-878.

Smuckler, R.H., Berg, R.J., & Gordon, D.F. (1988). *New Challenges New Opportunities: U.S. Cooperation for International Growth and Development in the 1990s.* Michigan State University, Center for Advanced Study of International Development, East Lansing, MI.

Snow, C.P. (1962). *Science and Government.* President and Fellows of Harvard College, Cambridge, MA.

Solutions. (1989). *Solutions.* Sept./Oct.:24.

Spring, A. & Hansen, A. (1985). The underside of development: agricultural development and women in Zambia. *Agriculture and Human Values.* 2 (1):60-67.

Stone, C. (1972). Should trees have standing? toward legal rights for natural objects. *Southern California Law Review.* 45 (Spring):450-501.

Stone, C. (1974). *Should Trees Have Standing? Toward Legal Rights for Natural Objects.* William Kaufman, San Francisco.

Stone, C. (1987). *Earth and Other Ethics: The Case for Moral Pluralism.* Harper and Row, New York.

Stout, B.A. & Nehring, R.F. (1988). Agriculture energy: 1988 and the future. *National Forum, The Phi Kappa Phi Journal.* Louisiana State University, Baton Rouge, LA. Summer.

Strange, M. (1988). *Family Farming: A New Economic Vision.* University of Nebraska Press, Lincoln, NB.

Summers, G., ed. (1983). *Technology and Social Change in Rural Areas.* Westview Press, Boulder, CO.

Suppe, F. (1987). The limited applicability of agricultural research. *Agriculture and Human Values*. Fall:4-13.

Swaminathan, M.S. (1988). Seeds and property rights: a view from the CGIAR system. *Seeds and Sovereignty: The Use and Control of Plant Genetic Resources*. (Kloppenburg, J.R., Jr., ed.) 231-254. Duke University Press, Durham, NC.

Swanson, B.E. (1986). The contribution of U.S. universities to training foreign agricultural scientists. *The Agricultural Scientific Enterprise: A System in Transition*. (Busch, L. & Lacy, W.B., eds.) 229-238. Westview Press, Boulder, CO.

Thompson, P.B. (1986). The social goals of agriculture. *Agriculture and Human Values*. 3 (4):32-42.

Thompson, P.B. (1987). Agricultural biotechnology and the rhetoric of risk: some conceptual issues. *The Environmental Professional*. 7:316-326.

Thompson, P.B. (1988a). Of cabbages and kings. *Public Affairs Quarterly*. 2 (1):69-87.

Thompson, P.B. (1988b). Ethics in agricultural research. *Journal of Agricultural Ethics*. 1 (1):11-20.

Thompson, P.B. (1988c). Ethical dilemmas in agriculture: the need for recognition and resolution. *Agriculture and Human Values*. 5 (4):4-15.

Thompson, P.B. (1988d). The philosophical rationale for U.S. agricultural policy. *U.S. Agriculture in a Global Setting*. (Cotwiler, M.A., ed.) Resources for the Future, Washington, DC.

Thrupp, L.A. (1989). Legitimizing local knowledge: from displacement to empowerment for third world people. *Agriculture and Human Values*. 4 (3)13-24.

Tweeten, L. (1983). Food for people and profit: ethics and capitalism. *The Farm and Food System in Transition - Emerging Policy Issues, No. FS5*. Cooperative Extension Service, Michigan State University.

Tweeten, L. (1984). *Causes and Consequences of Structural Change in the Farming Industry*. National Planning Association, Washington, DC.

Tweeten, L. (1987). Has the family farm been treated unjustly? *Is There a Moral Obligation to Save the Family Farm?* (Comstock, G., ed.) Iowa State University Press, Ames, IA.

U.S. Public Health Service. (1980). *Smoking and Health: A Report of the Surgeon General*. U.S. Department of Health, Education and Welfare; Public Health Service; Washington, DC; HEW Publication No. (PHS 79-50006).

Unger, S.H. (1982). *Controlling Technology: Ethics and the Responsible Engineer*. Holt, Rinehart and Winston, New York.

Uphoff, N. (1986). *Local Institutional Development: An Analytical Source Book with Cases*. Kumarian Press, West Hartford, CT.

USDA. (1975). *That We May Eat, The 1975 Yearbook of Agriculture*. 94th Congress, 1st Session House Document No 94-4.

Valliantos, E.G. (1977). *Fear in the Countryside: The Control of Agricultural Resources in Poor Countries*. Ballinger, Cambridge, MA.

Van den Bosch, R. (1978). *The Pesticide Conspiracy*. Doubleday, Garden City, NY.

VanDeVeer, D. & Pierce, C., ed. (1986). *People, Penguins, and Plastic Trees: Basic Issues in Environmental Ethics.* Wadsworth Publishing Company, Belmont, CA.

Vogeler, I. (1981). *The Myth of the Family Farm: Agribusiness Dominance of U.S. Agriculture.* Westview Press, Boulder, CO.

Wellhause, E.J. (1976). The agriculture of Mexico. *Scientific American.* 235:128-150.

White, L., Jr. (1967). The historical roots of our ecologic crisis. *Science.* 155:1203-1207.

White, L., Jr. (1968). *Machina Ex Deo: Essays in the Dynamism of Western Culture.* The MIT Press, Cambridge, MA.

Wilson, K. & Morren, G.E.B., Jr. (1990). *Systems Approaches for Improvement in Agriculture and Resource Management.* Macmillan Publishing Company, New York.

Wofsy, L. (1986). Biotechnology and the university. *Journal of Higher Education.* 57:477-492.

Wojcik, J. (1984). The American wisdom literature of farming. *Agriculture and Human Values.* 1 (4):26-38.

Wojcik, J. (1989). *The Arguments of Agriculture: A Casebook in Contemporary Agricultural Controversy.* Purdue University Press, West Lafayette, IN.

Wood, D. (1988). Crop germplasm: common heritage or farmers' heritage? *Seeds and Sovereignty: The Use and Control of Plant Genetic Resources.* (Kloppenburg, J.R., Jr., ed.) 274-289. Duke University Press, Durham, NC.

Woodwell, G.M. (1985). On the limits of nature. *The Global Possible.* (Repetto, R., ed.) Washington, D.C.

Woodwell, G.M. & Smith, H.H. (1969). *Diversity and Stability in Ecological Systems.* Brookhaven National Laboratory Publication No. 22. Clearinghouse for Federal and Technical Information, Springfield, VA.

World Commission on Environment and Development (1988). *Our Common Future.* Oxford University Press, Oxford.

World Bank. (1986). Poverty and hunger: issues and options for food security in developing countries. *A World Bank Policy Study.* Washington, DC.

Worster, D. (1985). A sense of soil: agricultural conservation and American culture. *Agriculture and Human Values.* 2 (4):36-40.

Wunderlich, G., ed. (1987). *Ideas, Facts and Choices: Three Lectures on Philosophy and Applied Economics.* Natural Resource Economics Division, Economic Research Service, U.S. Department of Agriculture, Staff Report AGES870512.

Yongberg, I.G. (1987). Moving from yesterday's agricultural technology: alternative farming systems in perspective. *Public Policy and Agricultural Research.* (Hadwiger, D.F. & Browne, W.P., eds.) St. Martin's Press, New York.

Zelenka, A. (1989). Land Stewardship Project, January.

Ziman, J. (1968). *Public Knowledge: The Social Dimension of Science.* Cambridge University Press, Cambridge.

About the Contributors

Miguel A. Altieri is associate professor and associate entomologist at the University of California-Berkeley. His research interests include the ecological features of traditional and modern agroecosystems.

Jeffrey Burkhardt is a philosopher in Ethics and Policy Studies in Agriculture and Natural Resources at the University of Florida.

Vernon B. Cardwell, professor in the Department of Agronomy and Plant Genetics at the University of Minnesota, is co-leader of a 3-year project investigating rural community development through agriculture.

Billie R. DeWalt is professor and chair of the Department of Anthropology at the University of Kentucky. He has published many books and articles on culture, agriculture, and development in Latin America.

G. Lain Ellis is a Ph.D. student in archeology at Texas A&M University. He is interested in the agricultures of extinct societies.

Merle L. Esmay, a former professor of agricultural engineering at Michigan State University, specialized in international agriculture until his death in April 1990.

David R. MacKenzie is director of the National Biological Impact Assessment Program, Cooperative State Research Service, U.S. Department of Agriculture.

Bryan G. Norton is professor of philosophy at Georgia Institute of Technology. His book, *Why Preserve Natural Variety?* was published by Princeton University Press.

Vernon W. Ruttan is Regents Professor of Agricultural and Applied Economics at the University of Minnesota and a member of the National Academy of Sciences.

Nancy W. Shankle is a specialist in bibliographic research. She is assistant professor of English at Abilene Christian University in Abilene, Texas.

Bill A. Stout is professor of agricultural engineering at Texas A&M University. He has been co-principal investigator for this NSF project. His most recent publication is *Handbook of Energy for World Agriculture.*

Paul B. Thompson, associate professor of philosophy and agricultural economics at Texas A&M University, is co-principal investigator for this NSF project. His publications include various articles and book chapters on agricultural ethics and public policy.

Deborah A. Tolman has an M.S. in horticulture from Virginia Polytechnic Institute. She is pursuing graduate study in agricultural economics at Texas A&M University.

Gary E. Varner has taught environmental ethics at Washington University of St. Louis and at the University of Wisconsin-Stevens Point. He is a visiting assistant professor of philosophy at Texas A&M University.

Index

Printed and bound by CPI Group (UK) Ltd, Croydon, CR0 4YY

23/10/2024

01778240-0001